Making Sense of Life

MAKING SENSE OF LIFE

Explaining Biological Development with Models, Metaphors, and Machines

EVELYN FOX KELLER

HARVARD UNIVERSITY PRESS
Cambridge, Massachusetts, and London, England 2002

Printed in the United States of America

Library of Congress Cataloging-in-Publication Data

Keller, Evelyn Fox, 1936–
Making sense of life : explaining biological
development with models, metaphors,
and machines / Evelyn Fox Keller.
p. cm.
Includes bibliographical references (p.).
ISBN 0-674-00746-8 (alk. paper)
1. Developmental biology. I. Title.

QH491 .K387 2002
570′.1—dc21 2001051559

Designed by Gwen Nefsky Frankfeldt

Contents

Preface

Thinking about how this book came to be is a bit like thinking about the book's core question: How does an organism come to be? Both are questions of origins, regarding a process or phenomenon that has no distinct point of origin and for which the dynamics of development are so complex as to resist the very possibility of a definitive answer. To be sure, it is possible to identify a defining moment in each case when all the various strands of prior history fuse into a single entity. For the organism, that moment comes with fertilization, and for a book, with the signing of a contract. But it can also be said that there is a certain arbitrariness in such an assignment, for in both cases decisive formative influences need to be sought as much in the prior histories of the respective contributing strands as in the moment of fusion. So too, how we think about such questions depends on the predilections we bring to an inquiry, on our presuppositions about what will count as an answer, on our explanatory preferences.

My own predilection is to look to history, and even to biography. And to the extent that books are products of personal history, my inclination is to look to personal biography, to the various dimensions of the historical trajectory that lies behind any project, and, in this case, that led to the writing of this particular book. For example, twenty years ago, trying to make sense of the communication gap that Barbara McClintock encountered in attempting to present her work on transposition, I found myself wondering about the explanatory preferences that scientists bring to their work. More specifically, I sought an understanding of why McClintock remained so deeply unsatisfied by efforts to explain biological development in terms of the central dogma of early molecular biology—efforts that many others clearly did find satisfying. Might I then claim this as the moment of origin for the present project? Possibly, but surely only as part-origin, for I can also identify both earlier and more recent roots of this preoccupation.

Some of the earlier roots date back to my very first interactions with students and workers in the life sciences. Originally trained in theoretical physics, I was introduced to biology in stages. In the first period, as a physicist-turned-molecular-biologist, initiated into the field by others of similar background, I scarcely encountered anyone who had started out in life science, and the only cultural novelty that was immediately visible was that evoked by the transition to experimental work. In the settings in which I then worked, the relation between theoretical and experimental work was a familiar one. The primary function of experiment was to test theoretical models. In other words, experiment, though crucial, remained secondary to theory. Only later, while working as a mathematical biologist and teaching a course to medical students on the uses of mathematical methods in biology, did I get my first glimpse of a more fundamental divide. This came at the end of a class on dimensional analysis—a remarkably simple method for reducing the number of independent variables, which requires nothing more than identifying the dimensions of the different variables. After introducing a biological problem described in terms of eleven variables, I used dimensional analysis to show that the relations among only three of these variables needed to be studied empirically; all other relations could be inferred logically. The students, however, were clearly unhappy: "But you haven't done the experiments," they complained. "So how can you *know?* How can you be sure?" That question stopped me in my tracks, and I have been thinking about it ever since.

I had been trained to see arguments based on mathematics and logic as determining, and experimental evidence as fallible. But others, I soon came to realize, regarded logical arguments as suspect. To them, experimental evidence, fallible as it might be, provided a far surer avenue to truth than did mathematical reasoning. And these others included not only students but also many members of the biological community. Their implicit question seemed to be: How could one know one's assumptions were correct? Where, in a purely deductive argument, was there room for the surprises that nature might have to offer, for mechanisms that might depart altogether from those imagined in our initial assumptions? Indeed, for some biologists, the gap between empirical and logical necessity loomed so large as to make the latter seem effectively irrelevant.

Explicitly, the response I encountered as a mathematical biologist in efforts to talk with experimental biologists about problems I saw as compelling was more commonly simply one indicating lack of interest, impatience, or even irritation. And over the years since, I have had ample opportunity to observe similar failures of communication virtually whenever experimental and mathematical biologists happened to be in the same room. Some of the difficulties are obvious. Disciplinary territoriality is one. Ignorance—on the one hand, of some of the most crucial experimental facts and, on the other, of basic mathematical techniques—is another. Also (although this became obvious to me only after I acquired greater distance from my first discipline), there was the tell-tale arrogance that seems to be so naturally imbibed with a training in theoretical physics, and the inevitable bristling such arrogance evokes among those trained differently.

But it soon became evident that beneath these obvious sources of irritation lay issues of substantially deeper import. As the student in my long-ago course asked: How can one actually know? Indeed, what is to count as knowledge? As explanation? As theory? Philosophers of science have traditionally tended to approach such questions in the abstract, as if they could be answered independently of historical or disciplinary context. My own experience, as both a scientist and an historian, suggests otherwise. It persuades me that answers to such questions are not given but contingent; not universal but rather matters of local, and historically specific, disciplinary culture. The communication gap that has persisted through most of this century between experimental and mathematical biologists provides especially conspicuous evidence of such cultural differences, and it constitutes a leitmotif that runs throughout this book. But once I was alerted to the problem, other variations in epistemological culture became evident as well, and they are both temporal and interdisciplinary. These reflect differences in questions, in available technology, in resources, and in cognitive, practical, and psychological needs. In other words, my interest here is not in what *should* count as an explanation in science but on what *does* count. And for this, one must look to the explanatory conventions operative at particular times and in particular areas of scientific research. Obviously, a huge and unwieldy task.

In the attempt to make the task somewhat more manageable, I chose to focus on the particular problem of embryonic development and on the kinds of accounts of how a fertilized egg develops into an adult organism that have been offered (and, at least by some, found satisfying) over the course of the twentieth century. This problem first attracted my attention many years ago while I was still working as a mathematical biologist. But ten years ago, having become aware of the dramatic renaissance this subject was undergoing among molecular biologists, I turned to it in earnest. As any biologist in the field will attest, it has been an extraordinarily exciting period. What makes the problem of development especially compelling, however, and particularly for this project, is both its difficulty and its historical resistance to any generally acceptable explanation in terms of either genetics or physico-chemical mechanisms. To be sure, there has been no shortage of attempts, but while many of these efforts did (or do) find adherents, at least for a time, no one of them has been able to claim lasting acceptance in the scientific community at large. Indeed, the very difficulty of the problem seems to have made it a particularly fertile arena for displaying diversity in epistemological values. Some might claim that only today do we finally have an explanation of embryonic development, but that claim has been made before, and, like other such claims, it reveals as much about its claimants as it does about its subject—as much about what is counting as an explanation in our time as about how an organism emerges from a fertilized egg.

The style of this work resembles that of a meditation, based on a wide range of examples. I make no effort to be comprehensive, and, taken as a whole, the book will strike some readers as idiosyncratic. And in many ways it is. Much of this idiosyncracy is due to my history.

Inevitably, I have incurred many debts in the course of writing this book. The Guggenheim Foundation and the Max Planck Institute for the History of Science have provided me with invaluable support, and the Rockefeller Foundation granted me idyllic accommodations at Bellagio for the final revisions of my manuscript. Thanks are also due to my editors at Harvard University Press, Susan Wallace Boehmer and Michael Fisher, for both their editorial suggestions and their sustaining confidence and encouragement; to Gregory Davis, George Homsy, Myra Jehlen, Jehane Kuhn, Ed Munro, and Lee Segel for their careful reading of individual chapters; to John Jungck and other reviewers of an early draft of the entire manuscript for a number of helpful suggestions; and to the many biologists who allowed me to consult with them about technical details and who patiently educated me and, in some cases, corrected my errors. But my largest debt is to Loup Verlet; without his friendship and critical reading of each chapter as it emerged, this book might never have seen the light of day. Necessarily, however, responsibility for any shortcomings or inadequacies remain mine alone.

Introduction

Paraphrasing the question Freud so famously asked about women, we might ask: What do biologists want? Is there an endpoint, a final goal, of biological inquiry? And if so, what would that endpoint be? In one sense, this question is no more meaningful than Freud's own: biologists surely come in as many different stripes as do women, harboring an equally disparate multitude of desires and goals. Yet, when asked about physicists (who vary fully as much as do biologists), the question does appear to make sense. What's more, in that case we seem to have a ready answer. Physicists, we are often told, seek to expand the boundaries of knowledge until nothing, at least nothing in the physical universe, is left unexplained: they want, in short, a "theory of everything." Furthermore, in that science we have a history of stunning examples of what such a comprehensive theory or explanation might look like. Think, for example, of Newton's theory of gravitation.

By comparison, the ambitions of biologists are manifestly less grandiose. They have no such exemplary grand theories to guide their aspirations, nor, indeed, do they seem to share such a concept of theory. Then what is it that they do want to know? Where is the common denominator that binds the epistemological aspirations of workers in the life sciences into a single disciplinary identity? At first glance, one might suppose the answer to be there, in the very concept of life—that is, we might say that the business of biologists is to make sense of life. But what exactly is meant by "making sense of life"? Throughout most of the past century, biologists have generally eschewed the possibility, or even the value, of an overarching theory of life. What then does count, and what has counted, as an explanation to workers in this discipline? Or, to rephrase the question somewhat, what is it that biological explanations have aimed at in the past, and what do they aim at today? How will we know when we have made sense of life?

An underlying thesis of this book is that no simple answer to these questions can be given. Typically, explanations in the biological sciences are provisional and partial, and the criteria by which they are judged are, and always have been, as heterogeneous as their subject matter. My aim is to illustrate this thesis. Just as the diversity of life, rather than its unity, has historically commanded the respect of life scientists, so too, I propose, the epistemological diversity of their aspirations demands our respect as historians and philosophers of science. No one can deny the extraordinary advances that have been made over the course of this past century in our understanding of vital processes. In fact, so dramatic have been these achievements that today, at the start of a new century, biology seems to be outflanking physics as the leading natural science.

Yet I would argue that, despite such unquestionable success, biology is scarcely any closer to a unified understanding (or theory) of the nature of life today than it was a hundred years ago. The models, metaphors, and machines that have contributed so much to our understanding provide neither unity nor completeness. They work to answer some questions while avoiding (even obscuring) others; they satisfy certain needs while failing to address others; in short, they leave the project of "making sense of life" with an essentially—and perhaps necessarily—mosaic structure.

Perhaps nowhere is the diversity of explanatory goals more conspicuous than in the history of efforts to answer the question: How are living entities formed? The question is itself ambiguous, sometimes referring to the emergence of life on earth and at other times to the unfolding of an individual life, that is, to the development of a complex multicellular organism from a fertilized egg. These two forms of the question of life's origins have not always been clearly distinguished and have, at least sometimes, been quite closely conjoined. The central focus of my project is not on biological science in general but on the form of the question that defines the subdisciplines that have been variously called morphogenesis, embryogenesis, and developmental biology. I ask what counts, and what has counted, as an "explanation" of biological development in individual organisms. To illustrate the diversity in answers that have been given to this question, I examine a wide range of models, metaphors, and machines that, over the past century, have helped biologists "explain" this process.

One form of explanation has come to dominate biological thought over the last few decades—the assumption that a catalogue of genes for an organism's traits will constitute an "understanding" of that organism. Yet, an increasing number of biologists are beginning to argue that no such catalogue—not even the sequence of the entire genome—can suffice to explain biological organization. The reason most commonly offered for their skepticism regarding genetic explanations is that the regulatory apparatus for turning genes on and off is distributed throughout the organism. In fact, debate between proponents of these two different views of explanation has raged throughout the history of genetics, and it is sometimes described as a debate between bottom-up and top-down (or reductionist and holist) strategies of explanation. But such differences, I argue, do not begin to exhaust the variability in what has counted over the course of this century—and what, for some, may even continue to count—as an explanation of biological development.

Much of the heterogeneity of criteria by which explanations are judged can be understood by attending to the local practices—the techniques, the instruments, and the experimental systems—of a particular scientific subculture. I claim, however, that reference to practice, at least in the most familiar senses of that word, does not quite suffice, that something more is needed. Techniques, instruments, and experimental systems are well known to be extraordinarily variable, but so too are the meanings attributed to so basic a term as "understanding." Because reference to scientific practices rarely encompasses that variability, and in order to underscore the dependence of explanatory criteria on the epistemic needs of a particular scientific subculture, I invoke the notion of epistemological culture, by which I mean the norms and mores of a particular group of scientists that underlie the particular meanings they give to words like theory, knowledge, explanation, and understanding, and even to the concept of practice itself.[1]

My project departs from virtually all of the existing philosophical literature on scientific explanation in several important ways. In contrast to the view most familiar to philosophers of science—namely, that explanatory adequacy (and/or explanatory power) is self-evident in science[2]—I approach the question of the meaning of explanation by asking: What counts as an explanation in actual scientific practice? In sympathy with Steven Weinberg's recommendation to philosophers of physics, I want to ground my discussion of explanation in biology in that which leads biologists to say *Aha!*[3] Posing the question in this way, however, one quickly finds a bewildering set of answers. And for me, this very variability commands historical and philosophical interest. Thus, rather than beginning with any presuppositions about what an explanation ought to be in order to qualify as satisfying, I start with the variety of explanations that biologists at different times, and in different contexts, have found to be useful. My approach might be regarded as empirical rather than philosophical (at least in the strict sense of that term), and hence as complementing rather than as supplementing more traditional work in the philosophy of explanation.

A description of a phenomenon counts as an explanation, I argue, if and only if it meets the needs of an individual or community. The challenge, therefore, is to understand the needs that different kinds of explanations meet. Needs do of course vary, and inevitably so: they vary not only with the state of the science at a particular time, with local technological, social, and economic opportunities, but also with larger cultural preoccupations. By focusing on the problems that arise in developmental biology, I try to exhibit the range of different needs that can come into play in shaping the criteria by which an explanation of a given phenomenon may be judged. More generally, I claim that the temporal, disciplinary, and cultural specificity of needs is responsible for the specificity of what I call an epistemological culture. If a successful explanation is one that must satisfy the local needs of a particular scientific culture, so too might we think of the adjustment of local meanings attributed to terms like theory, knowledge, and understanding as an effort to meet those same needs.

Needless to say, one might classify the explanatory needs of a scientific culture in many different ways. At the very early stages of this project, I attempted to categorize explanatory efforts according to their ability to satisfy needs for prediction, control, or narrative coherence. But that of course is only one possible taxonomy. Alternatively, I considered a categorization in terms of cognitive, instrumental, and social/psychological needs. Eventually, however, I abandoned all such attempts at taxonomy. I found that explanations typically aim at satisfying many different kinds of needs at once, with varying degrees of success, and that what most conspicuously distinguishes one effort from another is the relative urgency of particular needs to an individual or community. Furthermore, analysis of the literature on biological development over the course of the century reveals not only great variability in criteria—over time and between different research schools—for what might count as an explanation of biological development, but also how flexible these criteria can be. Thus, in the end, I opted for a simpler and more empirical approach to the organization of this book.

My discussion is divided into three parts. To highlight the diversity of explanatory efforts in the past century, I begin with examples of what would surely *not* count as explanation to most biologists working today but which nevertheless did count (at least for some) when they were first introduced. Part One ("Models: Explaining Development without the Help of Genes") describes and examines a number of efforts to explain the emergence of biological form from the first half of the century—prior to 1953—through the use of physical and mathematical models. In Chapter 1, I focus on Stéphane Leduc's early efforts in "synthetic biology" and "artificial life" (1905–1928); in Chapter 2, on D'Arcy Thompson's still widely read *On Growth and Form* (first published in 1917); and in Chapter 3, on Alan Turing's mathematical model of embryogenesis (in 1952) and its place in the troubled history of mathematical biology. Viewed in hindsight, from the standpoint of the great triumphs of molecular biology, none of these efforts would be regarded as successful. Indeed, in none of the examples I have chosen do genes make more than a passing appearance. Because they shed little if any light on the actual processes occurring in the development of biological form, they have been seen as largely irrelevant to the needs of experimentalists in the field. But all three, in varying degrees, did work toward the satisfaction of other scientific needs: they helped to fill a conceptual void formed not simply by an absence of concrete information about the actual processes of development but by an absence even of imaginable answers to the question of how complex forms *could,* in principle, emerge by the operation of purely physical principles.

In Part Two ("Metaphors: Genes and Developmental Narratives") I turn to the more familiar explanations of development that have come from genetics. Concepts like gene action and genetic program (Chapter 4), feedback (Chapter 5), and positional information (Chapter 6) have proven of enormous importance in constructing explanations of development out of genetic data. How, I ask, do these concepts contribute to explanatory satisfaction? My answer to this question requires that we attend closely to the meanings of the words we use and the ways in which we use them, and take seriously the linguistic and narrative dimensions of explanation. The core of my argument is that much of the theoretical work involved in constructing persuasive narratives of development out of genetic data depends on productive use of the cognitive tensions generated by ambiguity and polysemy and, more generally, by the introduction of novel metaphors.[4]

I conclude with a discussion of new modes of explanation that began to emerge at the end of the twentieth century and that promise to assume ever greater importance in the twenty-first. The third and last part of this book ("Machines: Understanding Development with Computers, Recombinant DNA, and Molecular Imaging") is concerned with the epistemic mutations brought about by the arrival of new machines. In these chapters, I explore some of the ways in which the dramatic technological developments of the last thirty years— especially in computers and recombinant DNA—have begun to transform biologists' understanding of what counts as an explanation. The emergence of visual technologies that provide direct access to the dynamics of living cells (the subject of Chapter 7) may recall an older tradition in the biological sciences in which the experience of direct seeing (or watching) was granted clear and often vociferous priority over theoretical speculation, but it also adds complications to the meaning of direct seeing that could not have been imagined in the earlier period. In Chapter 8 I turn to the ways in which new technological developments have undermined traditional resistances among experimental biologists to the role of theory and mathematics in their work; but here too, these developments have also changed the very meanings of the word theory. Techniques of recombinant DNA have unleashed a wealth of data so vast as to overwhelm the powers of intuitive reasoning, and biologists are increasingly coming to rely on computers to make sense of that data. Where one technology has generated a need and the other a resource, together they have created a new stage for mathematical and computational modeling in biology. But if, as some say, a new mathematical biology is emerging, it is an activity bearing little resemblance to the mathematical biology of earlier decades. Finally, in Chapter 9, I return to the question of artificial life, only this time around in the context of computer simulations. For whom, I ask, do these simulations have explanatory value, and why?

The range of issues to which these chapters take me is vast, and nowhere do I attempt to be exhaustive. Nor are the examples in each chapter intended to be representative. Rather, they are chosen to illustrate the breadth of explanatory styles that have come into play. Indeed, the book taken as a whole might be said to exhibit the same mosaic quality as my examples. Read it, then, more as an attempt to raise than to resolve questions, more as an effort to stimulate new kinds of inquiry than to settle old scores.

PART ONE

Models: Explaining Development without the Help of Genes

atson and Crick's elucidation of the structure of DNA in 1953 constituted a watershed of inestimable proportions in the history of biology. But here, in the first part of this book, I discuss several efforts to explain the emergence of biological form prior to this watershed, efforts that relied not on experimental analyses of biological systems, nor even on biological concepts, but rather on the use of physical and mathematical models employing only the concepts of physics and chemistry.

Stéphane Leduc's attempts to explain the emergence of biological form through the synthesis of artificial organisms out of inorganic chemicals (the subject of Chapter 1) can be of no interest to today's biologists; indeed, these attempts seem almost self-evidently absurd. Yet for a brief period in the first two decades of the twentieth century they attracted considerable interest in the scientific literature and were seen by many readers as promising to illuminate the nature and origin of life. How are we to make sense of the success, however short-lived, of a model or explanation that today seems so utterly ineffective?

My answer is simple: we need to recognize that the growth of biological understanding has depended on the formulation of models and explanations that answer some questions while ignoring others, satisfying only some of the many different kinds of needs which explanations are designed to meet. Leduc's models met a need that was widely felt in his own time, even if not in ours: they demonstrated that complex forms—comparable in complexity to those found in the living world—could be brought into existence by recognizable physical and chemical processes. In short, they demonstrated the possibility of such forms emerging without the intervention of extra-physical powers.

D'Arcy Thompson's *On Growth and Form,* the subject of Chapter 2, first appeared in the same period as Leduc's publications, and it was aimed at meeting a similar need. Like Leduc, Thompson took on the task of uprooting the remaining vestiges of vitalism—of showing that the principles of physics and chemistry could suffice, by themselves, to account for the growth and development of biological form. But unlike the writings of his contemporary, Thompson's work met with instant acclaim, and it went on to become a classic in the literature of biology. Yet there is little evidence that it has had any more influence on either the conceptual or experimental practices of working biologists than did the work of Leduc. How then are we to account for its extraordinary success? History has made an icon of D'Arcy Thompson and an embarrassment of Stéphane Leduc. Why? What needs, if not those of practicing biologists in their day-to-day work, was the work of the former, but not that of the latter, in fact able to meet? What needs does it meet today?

Chapter 3 addresses the history of mathematical biology, and my discussion of Alan Turing's 1952 model of embryogenesis is situated in the context of this history. Before turning to Turing, however, I briefly review the energetic but ultimately unsuccessful efforts of Nicolas Rashevsky to establish a discipline of mathematical biology during the 1930s and 1940s. Rashevsky explicitly credited D'Arcy Thompson's work as his inspiration for attempting to build "a systematic mathematical biology, similar in its structure and aims to mathematical physics,"[1] and in many ways Turing's contribution, although far more limited in scope, can also be seen as a continuation of Thompson's earlier efforts. But a study of the approaches of both Rashevsky and Turing to the problems of biology reveals with far greater clarity than does a study of Thompson's work the cultural divide (of which Thompson was clearly aware, and about which he had much to say) that has historically prevailed between the mathematical and biological sciences. From the juxtaposition of Rashevsky and Turing with the responses of experimental biologists, we can begin to see the lineaments of two clearly distinct epistemological cultures.

One of the principal markers of cultural difference can be found in the value accorded to models that are clearly acknowledged to be fictional, that make no pretense to realism. Examination of such differences is of interest in and of itself; more specifically, however, I suggest that doing so helps us to understand why mathematical biology, although by now a well-established discipline in its own right, has historically failed to find acceptance as a proper part of biology. Despite some indications that this may be changing (a subject I address in Chapter 8), the fact remains that none of these early efforts had a noticeable effect on what subsequently emerged as the dominant explanatory framework of twentieth-century biology. That explanation, as everyone who follows biology today is aware, is to be found in genetics, and it is striking that neither Leduc nor Thompson nor Rashevsky nor Turing considered it necessary to include genes in their models for development. From today's perspective, this fact alone locates all these early attempts in a different, and now superseded, historical era.

CHAPTER ONE

Synthetic Biology and the Origin of Living Form

There is, I think, no more wonderful and illuminating spectacle than that of an osmotic growth,—a crude lump of brute inanimate matter germinating before our very eyes, putting forth bud and stem and root and branch and leaf and fruit, with no stimulus from germ or seed, without even the presence of organic matter. For these mineral growths are not mere crystallizations as many suppose . . . They imitate the forms, the colour, the texture, and even the microscopical structure of organic growth so closely as to deceive the very elect.

W. Deane Butcher, preface to Stéphane Leduc, *The Mechanism of Life* (1911)

ompared with physics, biology is a young science. The word itself was introduced only in 1802 (most famously by Jean-Baptiste Lamarck, but independently and in the same year by Gottfried Reinhold Treviranus and Lorenz Oken in Germany) to designate a new "science of life." Such a designation was in turn prompted by Lamarck's advocacy of a new ontology—one that emphasized the commonality of the forms of animal and plant life and stressed their distinctiveness in relation to the "non-living."[1] But what was the significance of demarcating biology as a distinctive scientific discipline, with its own distinctive subject matter? One consequence, I suggest, was the establishment of an intellectual space from which the category of "life" could be taken as a given, as a domain of natural phenomena declared to be itself "natural," and hence one to be investigated without calling the limits of that category into question. From within this space, the manifold and largely mysterious properties of living beings offered challenge enough, and no need was felt for a definition or theory of "life." Not only was there no need to answer the question "What is life?" but the question itself would have had little grip. Life was manifest, and the goal of biology was simply to investigate, and to make sense of, the ways and means of its manifestations.

Yet, even while the new disciplinary demarcation might offer some students of living phenomena a refuge from the need to say what life *is,* elsewhere, that same demarcation also, and simultaneously, worked to exacerbate that very need. Thus, it was inevitable that the demarcation of biology as a separate science generated a compelling and enduring tension that ultimately came to focus precisely on the boundary that delimited the category of living beings, and on the violability or inviolability of that boundary. By the beginning of the twentieth century, the question of what life is was being asked with increasing urgency, and, along with such companion figures as the "riddle of life," or "secret of life," was understood increasingly as provocation, as demanding an answer, a solution, an unmasking. Is the demarcation between the living and the non-living finally so categorical as to admit no intermediates, no bridges that might link the two domains? Would not the very possibility of a scientific account of the origins of life require such bridges? Also, and in much the same spirit, just how categorical is the demarcation between the discipline of biology and those of physics and chemistry? Can this boundary too not be bridged or, even better, in fact breached?

Lamarck's own ambivalence on these questions was manifest. Standing at the disciplinary fork that he himself had urged, though he would not live to see institutionalized, his *Zoological Philosophy* (written in 1809) was, from first to last, "an enquiry into the physical causes which give rise to the phenomena of life."[2] He wrote, "Nature has no need for special laws, those which generally control all bodies are perfectly sufficient for the purpose." However, "if we wish to arrive at a real knowledge of . . . what are the causes and laws which control so wonderful a natural phenomenon, and how life itself can originate those numerous and astonishing phenomena exhibited by living bodies, we must above all pay very close attention to the differences existing between inorganic and living bodies" (p. 191). In doing so, Lamarck finds that "between crude or inorganic bodies and living bodies there exists an immense difference, a great hiatus, in short, a radical distinction such that no inorganic body whatever can even be approached by the simplest of living bodies" (p. 194).

Even so, he nonetheless concludes that, at the very lowest levels of the scales of animal and vegetable organization, "Nature . . . herself creates the rudiments of organisation in masses where it did not previously exist" (p. 236). Spontaneous generation would provide the missing link. Thus, even in Lamarck's own writings, his demarcation could be read at one and the same time as offering assurance of the autonomy of biology and its subject, and as provocation, goading efforts to undermine that autonomy. And so it has indeed been read ever since.

My focus in this chapter is not on the immediate legacy of Lamarck's demarcation but on its status one hundred years later, when evidence of the more provocative aspects of that legacy had become so conspicuous. The nineteenth century, despite persistent efforts on the part of biologists throughout, had failed to yield a satisfying definition of life. For every proposed list of essential properties, either an exception in the living world could be found or a qualifying candidate among the manifestly nonliving. Yet, in spite of (or perhaps even exacerbated by) this failure, the question "What is life?"—if we can judge by the frequency of its iteration—had come to take on some urgency by the early decades of the twentieth century. But how, if not by definition, might one try to answer this question? How else might one go about solving the riddle of life?

To many authors writing in the early part of the twentieth century, the alternative seemed obvious: the question of what life is was to be answered not by induction but by production, not by analysis but by synthesis. Jacques Loeb was one of many to whom it seemed self-evident that the route to understanding the nature of life lay in producing life in the laboratory. In a lecture he gave in Hamburg in 1911 (subsequently published in *Popular Science Monthly* and reprinted in *The Mechanistic Conception of Life,* 1912), he argued that "we must either succeed in producing living matter artificially, or we must find the reasons why this is impossible." Furthermore, he was confident we would succeed: "By the 'riddle of life' not everybody will understand the same thing. We all, however, desire to know how life originates . . . We are not yet able to give an answer to the question as to how life originated on the earth . . . Nothing indicates, however, at present that the artificial production of living matter is beyond the possibilities of science."[3] In these few sentences, Loeb articulates a link that was common in, if not fundamental to, the logic assumed by many biologists of his time, namely, the conjunction of questions about the nature of life with those about its origin. Such an association was of course already present in the writings of Lamarck, but in Loeb's time the habit of equating understanding with construction would have effectively eliminated any perceptible difference between the questions.[4]

Making Life: Simulations and Realizations

Although Loeb was convinced that the artificial production of life not only would tell us how life originated on earth but would in fact constitute the test of an adequate explanation, he himself did not attempt the synthesis of life *de novo.* Rather, his strategy for addressing the riddle of life was to pursue what he saw as the more modest goal of understanding the riddle of individual life, and it was to this end that he sought to initiate (if not actually produce) the life of individual organisms by artificial means.[5] However, a number of other scientists—in Germany, France, Switzerland, England, and Mexico—were notably less reticent.[6] Though they shared with Loeb the recognition that the actual production of living matter from manifestly non-living components was not yet possible, they displayed a remarkable confidence that that goal could be approached incrementally, by the simulation of ever more lifelike constructions. Indeed, where Loeb saw the production of individual life by artificial means as an approach to the more general problem of the artificial production of life on earth, they saw the proper route to that goal in the production of what they themselves called artificial organisms.

So widespread were such efforts in the latter part of the nineteenth century and the first two decades of the twentieth that a new term seemed called for as a designator of a new science. Some used "synthetic biology"; others, "plasmogeny."[7] Notably, both terms contained the same ambiguity, eliding in their very definition what is, at least to us, the conspicuous difference between the production of artificial life and the artificial production of life. Indeed, Ernst Haeckel had defined plasmogonie in 1868 as the spontaneous generation of an organism out of "an organic formative fluid,"[8] but fifty years later, a number of authors adapted the term to designate their own efforts at producing lifelike forms out of inorganic fluids and crystals. In a retrospective summary of the work that he and others had been engaged in for several decades, the Mexican physiologist A. L. Herrera took the liberty of extending the history of this new discipline even further back in time, but he was particularly enthusiastic about its recent progress: "Plasmogeny, originated by Nollet, who discovered osmosis in 1748, was a child about 1885, with the work of Bütschli and Quincke. Today it is an adult in possession of its full strength and faculties. Who knows when it will reach its objective, which is the synthesis of living matter?"[9] More generally, however, and especially in the semipopular press, these productions were simply referred to as artificial organisms or even, as artificial life.

Contemporary readers will no doubt be more familiar with the usage of the term artificial life in connection with the current work of Christopher Langton and his colleagues on computer simulations of life. As Langton explains in his introductory article, "Artificial Life is the study of man-made systems that exhibit behaviors characteristic of natural living systems."[10] My use of the term in the present discussion is not entirely anachronistic. Almost a century before Langton, the same term was in fact widely employed, albeit more expansively, to describe a range of ventures. How much overlap can be found between these earlier ventures and contemporary efforts? In Chapter 9, I will try to make the case for a number of striking similarities; but for now, what is most immediately likely to impress us are the differences.

A useful introduction to the earlier usage is provided by a two-part overview of the subject that appeared in *Scientific American* in 1911 under the title "The Creation of 'Artificial Life.'"[11] Benjamin C. Gruenberg, a biologist at the National Institutes of Health, explains the need to divide his review into two parts by calling attention to, and attempting to distinguish, two different ways, corresponding to "two distinctive phases of scientific research," in which the term was then being used. As he notes:

> The expression "artificial life" has been used for two entirely different sets of ideas. On the one hand is the attempt to make artificial combinations of matter behave like living protoplasm—that is, to make live matter out of the non-living materials lying all about us. On the other hand is the attempt to make the eggs of various animals develop without the co-operation of the sperm—or to produce "artificial parthenogenesis," as it has been called. Both kinds of experiments are calculated to throw much light on the fundamental

> nature of "life"; but they differ considerably in their methods as well as in the point of view that prompts them.[12]

Accordingly, he would begin in Part I by discussing efforts to create "artificial life" in the first and rather bolder (or, as he put it, the "cruder") sense of "The Making of Living Matter from Nonliving," deferring his discussion of the other "phase of research," namely, the project of Loeb and others of artificially inducing the onset of life, to Part II.[13]

I will adhere to Gruenberg's distinction, confining my discussion, as he did in the first part of his review, to the early work on artificial life in the first sense of the term, and for two reasons. First, this meaning bears the closer affinity with current usage. And second, this sense of the term is, as Gruenberg notes, "connected intimately with [the problem] of the origin of life, and also with that of the characteristics or distinctive properties of living matter."[14] Yet it is worth pausing to ask: Why would such manifestly distinct efforts—one resulting in a structure which (at least to us) is so unambiguously an artifact, however closely it might appear to resemble a living organism, and the other, in a body that is equally unambiguously a living organism—ever have been joined under a single term in the first place? Should we read this conjoining as indicative of nothing more than a laxity of language?

I think not. Rather, I suggest that such a conjunction reveals two loci of uncertainty that will prove relevant to the perceived value of both kinds of research at the time they were being conducted. On the one hand, it reflects an uncertainty, already indicated in Loeb's remarks, over the epistemological proximity between the origins of individual life and the origins of life on earth, and of the value an explanation of the former would have for our understanding of the latter.[15] On the other hand, and more importantly for the purposes of this essay, it reflects an uncertainty over the proximity of these constructions to real organisms (an uncertainty already apparent in the appropriation of Haeckel's term plasmogonie, originally defined as the spontaneous generation of an organism, as well as in the ambiguity inherent in the term "synthetic," meaning both constructed and artificial). At least part of the perceived value of such constructions, particularly in the context of the spontaneous generation debates, lay precisely in their ambiguous (and indeed liminal) status—that is, in their resistance to decisive location in either realm, the living or the non-living.

The Mechanism of Life, Spontaneous Generation, and Osmotic Growths

"It is a marvelous spectacle," wrote Stéphane Leduc, "to see a formless fragment of calcium salt grow into a shell, a madrepore, or a fungus, and this as the result of a simple physical force."[16] Writing a quarter century later, Thomas Mann concurred:

> I shall never forget the sight. The vessel of crystallization was three-quarters full of slightly muddy water—that is, dilute water-glass—and from the sandy bottom there strove upwards a grotesque little landscape of variously coloured growths: a confused vegetation of blue, green, and brown shoots which reminded one of algae, mushrooms, attached polyps, also moss, then mussels, fruit pods, little trees or twigs from trees, here and there of limbs. It was the most remarkable sight I ever saw, and remarkable not so much for its appearance, strange and amazing though that was, as on account of its profoundly melancholy nature. For when Father Leverkühn asked us what we thought of it and we timidly answered him that they might be plants: "No," he replied, "they are not, they only act that way. But do not think the less of them. Precisely because they do, because they try to as hard as they can, they are worthy of all respect."
>
> It turned out that these growths were entirely unorganic in their origin; they existed by virtue of chemicals from the apothecary's

> shop, the "Blessed Messengers." Before pouring the water-glass, Jonathan had sprinkled the sand at the bottom with various crystals; if I mistake not potassium chromate and sulphate of copper. From this sowing, as the result of a physical process called "osmotic pressure," there sprang the pathetic crop for which their producer at once and urgently claimed our sympathy. He showed us that these pathetic imitations of life were light-seeking, helio-tropic, as science calls it. He exposed the aquarium to the sun-light, shading three sides against it, and behold, toward that one pane through which the light fell, thither straightway slanted the whole equivocal kith and kin: mushrooms, phallic polyp-stalks, little trees, algae, half-formed limbs. Indeed, they so yearned after warmth and joy that they actually clung to the pane and stuck fast there.
>
> "And even so they are dead," said Jonathan, and tears came in his eyes, while Adrian, as of course I saw, was shaken with suppressed laughter. For my part, I must leave it to the reader's judgment whether that sort of thing is matter for laughter or tears.[17]

This excerpt from Mann points to the persistence of popular reverberations of Leduc's work on diffusion and osmotic growths over several decades. Furthermore, Leduc was widely recognized among scientists of his generation as an eminent member of the medical faculty at the University of Nantes who had devoted much of his professional career to studies in biophysics. Yet, apart from D'Arcy Thompson's use of his work (see Chapter 2) and apart from the few remaining copies of his published books, little trace of Leduc's scientific life survives. As a contribution to the history of science, his efforts amounted to little more than a blip on the screen, and an eccentric one at that. John Farley, for example, in his review of the history of the spontaneous generation controversy, refers to the "strange experiments" of Leduc and his Mexican colleague, Alfonso Herrera, and writes, "They seem to verge on the absurd."[18] But from my perspective, this is just what makes them interesting. As an episode in the history of biological explanation, the ambitions those efforts reflected, as well as the interest they evoked in their time, are illuminating precisely in proportion to what may now appear to us as their absurdity.

The original title of Leduc's 1910 treatise *(Physico-Chemical Theory of Life and Spontaneous Generation)* describes his interests well. It followed two earlier books and was succeeded two years later by another.[19] All of these were based on his experimental efforts to close the gap between the living and non-living through the reconstruction, by purely physical means, of phenomena heretofore seen as inherently biological, and the "fabrication of forms resembling those of the lowest organisms."[20] In Leduc's view, the standard "course of development of every branch of natural science" begins with "observation and classification." "The next step," however, "is to decompose the more complex phenomena in order to determine the physical mechanism underlying them." "Finally," he writes, "when the mechanism of a phenomenon is understood, it becomes possible to reproduce it, to repeat it by directing the physical forces which are its cause—the science has now become synthetical" (p. 113). Yet his own work in synthetic biology suggests a manifest reversal of the second and final stages: like Loeb, he sought to understand the mechanisms of living phenomena by proceeding directly to the final, or synthetical, stage.

His earliest efforts were directed toward a physical reenactment of that most fundamental of biological phenomena, karyokinesis (mitotic division)—first described in 1876 by Hermann Fol. And indeed, in a presentation to the French Academy of Sciences meeting in Grenoble in 1904, he claimed the ability "to produce by diffusion not only the achromatic spindle, but also the segmentation of the chromatin, and the division of the nucleus" (p. 92). Here is his description of his procedure:

> We cover a perfectly horizontal glass plate with a semi-saturated solution of potassium nitrate to represent the cytoplasm of the cell.

> The nucleus in the centre is reproduced by a drop of the same solution coloured by a trace of Indian ink, the solid particles of which will represent the chromatin granules of the nucleus. The addition of the Indian ink will have slightly lowered the concentration of the central drop, and this is in accordance with nature, since the osmotic pressure of the nucleus is somewhat less than that of the plasma. We next place on either side of the drop which represents the nucleus a coloured drop of solution more concentrated than the cytoplasm solution. The particles of Indian ink in the central drop arrange themselves in a long coloured ribbon, having a beaded appearance. (p. 93)

Observing the successive phases of "artificial karyokinesis" unfold before his very eyes, he concluded, "The resemblance of these successive phenomena to those of natural karyokinesis is of the closest. The experiment shows that diffusion is quite sufficient to produce organic karyokinesis, and that the only physical force required is that of osmotic pressure" (p. 94; see Figure 1).

Leduc's parallel project, the synthesis of "artificial organisms," focused on the cellular rather than the nuclear level, and he also presented his first successes in this effort at the meeting in Grenoble. Here, his principal technique derived from an earlier demonstration by Moriz Traube in 1867 showing that "artificial cells" could be generated from the osmotic properties of chemical precipitates.[21] As he explained, "When a soluble substance in concentrated solution is immersed in a liquid which forms with it a colloidal precipitate, its surface becomes encased in a thin layer of precipitate which gradually forms an osmotic membrane round it" (p. 94). Furthermore, increase in osmotic pressure gives rise to "osmotic growth," either in the volume of the original "cell" or in the production of several cells. Leduc describes the effect as follows: "The first cell gives birth to a second cell or vesicle, and this to a third, and so on, so that we finally obtain an as-

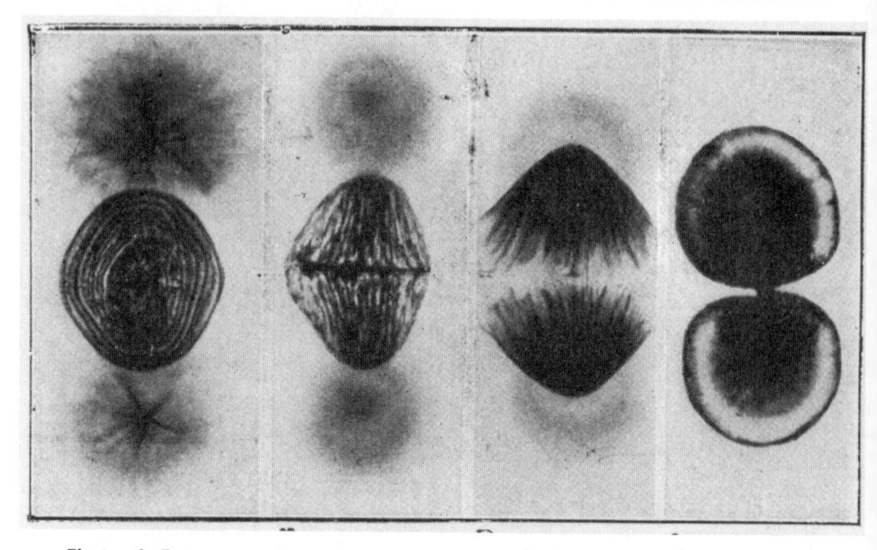

Figure 1. Four successive stages in the production of artificial karyokinesis by diffusion. (Leduc, *The Mechanism of Life,* fig. 32.)

sociation of microscopic cellular cavities, separated by osmotic walls—a structure completely analogous to that which we meet with in a living organism" (p. 134).

Indeed, by employing a variety of metallic salts and alkaline silicates (for example, ferrocyanide of copper, potash, and sodium phosphate) and adjusting their proportions and the stage of "growth" at which they were added, Leduc was able to produce a number of spectacular effects—inorganic structures exhibiting a quite dramatic similitude to the growth and form of ordinary vegetable and marine life (see Figure 2). By "appropriate means," it proved possible to produce "terminal organs resembling flowers and seed-capsules," "corral-like forms," and "remarkable fungus-like forms." Also,

> Shell-like osmotic productions may be grown by sowing the mineral in a very shallow layer of concentrated solution, a centimetre or less

> in depth, and pouring over this a less concentrated layer in solution. By varying the solution or concentration we may thus grow an infinite variety of shell forms . . . With salts of manganese, the chloride, citrate or sulphate, the stages of evolution of the growth are distinguished not only by diversities of form, but also by modifications of colour . . . Very beautiful growths may be obtained by sowing calcium chloride in a solution of potassium carbonate, with the addition of 2 per cent. of a saturated solution of tribasic potassium phosphate. This will give capsules with figured belts, vertical lines at regular intervals, or transverse stripes composed of projecting dots such as may be seen in many sea-urchins.

Moreover, some of Leduc's osmotic "organisms" exhibited "a considerable amount of mobility," and even (although his still photographs could scarcely attest to such a claim) an apparent capacity for "free-swimming." Lastly, and perhaps of greatest importance to most readers, these osmotic growths (like the original

Figure 2. Osmotic productions. (Leduc, *The Mechanism of Life,* frontispiece.)

osmotic "cells") appeared capable of reproducing: "Frequently a single seed or stock will give rise to a whole series of osmotic growths. A vesicle is first produced, and then a contraction appears around the vesicle, and this contraction increases till a portion of the vesicle is cut off and swims away free like an amoeba."[22]

In sum, then, Leduc's osmotic growths exhibited all the essential properties conventionally attributed to living organisms: growth and reproduction; assimilation and elimination ("Osmotic growths absorb material from the medium in which they grow, submit it to chemical metamorphosis, and eject the waste products of the reaction into the surrounding medium"); and morphogenesis, the generation of the particular and characteristic form which Leduc regarded as "the essential character of the living being." Indeed, he viewed the entire ensemble of functions that constitute life as itself "conditioned by form, that is, the external, internal, and molecular forms of the living being." So extensive and so close were the resemblances he was able to observe that Leduc persuaded himself that here, in the forces of osmotic pressure, he had found the fundamental physical basis (or cause) of all of life's most essential properties.[23] As he concludes in his final paragraph, "When we see under our own eyes the cells of calcium become organized, develop and grow in close imitation of the forms of life, we cannot doubt that such a transformation has often occurred in the past history of our planet, and the conviction becomes irresistible that osmosis has played a predominant role in the history of our earth and its inhabitants. It is a matter of astonishment that the scientist has taken no notice of the active part which osmosis has played in the evolution of our earth" (pp. 171–172).

Thomas Mann left to his readers' judgment the question of whether to laugh or cry. But what about Leduc's readers? How did they respond?

Leduc's Reception

Leduc was far from alone either in his ambition to close the gap between the living and non-living through experimental invention or in his particular strategy for achieving this ambition. Why, then, single him out for special attention? Because, quite simply, he was chosen by the readers of his day to represent these efforts. In the years between 1905 and 1913, Leduc's efforts received the lion's share of publicity in the scientific press, and especially in England and the United States.[24] From some of these readers, Leduc's achievements did indeed evoke a bit of laughter (or, more accurately, mockery), but for a surprising number of others they were seen as a matter for neither laughter nor tears but for interest, for possibility, and even for "proof" of fundamental propositions about the nature of life.

One of the earliest responses came from Leduc's colleague and old friend from Vienna, Moriz Benedikt. Benedikt was extremely positive. He wrote that Leduc's experiments open an entirely new view on the origins of form and life. They are "distinguished by their simplicity and their clarity. The new phase which they inaugurate deserves to be followed with the greatest attention." Especially interesting, thought Benedikt, and "the great triumph of Leduc's researches is the production, by the physical forces of diffusion, of the *figure* of karyokinesis, and the movements of nuclear and cell division."[25] But it was not just Leduc's friends who believed his work merited close attention. In the same month, a report in *Scientific American* reviewed in some detail his "highly interesting experiments, where the germination and growth of the natural cells was reproduced in artificial cells."[26] One year later, a second report appeared in the *Supplement* of the same journal in which the Paris correspondent described "a phenomenon hitherto unobserved . . . [that] has a striking analogy with the segmentation of the yolk of the egg during incubation." This phenomenon, brought to notice by "M. Stéphane Leduc, the eminent French physicist [sic]," leads to "a most important conclusion": it enables us "to produce the different kinds of structures used in the formation of living animals," under experimental conditions "realized in all the natural waters, and especially in the present and past state of the seas."[27]

Alfred Gradenwitz, writing in *Scientific American* in March 1907, was somewhat more ambivalent. But Gradenwitz's reservations had to do with Leduc's claims to novelty rather than with the work itself.[28] Indeed, insofar as it contributed to "the discovery of transitional stages between inert matter and living beings," the value of that work was unmistakable to Gradenwitz.[29] "Prof. Leduc," he writes, "has found the vital functions in animal and vegetable cells to be controlled exclusively by the physical laws of diffusion (osmosis) and cohesion (molecular attraction). On the basis of these phenomena he has even succeeded in artificially producing objects which, not only in appearance but in behavior, closely resemble natural cells growing, absorbing food, and propagating themselves in exactly the same way." "Nevertheless," he continues, "they are not living beings of any sort, but artificial bodies formed in the laboratory of the chemist."

What then can they tell us about actual living beings? Quite a lot, it seems: they "prove that the fundamental element of animal and vegetable organisms, viz. the cell, is exclusively controlled in its vital functions by the same physical laws that govern the forms of the mineral kingdom." In this way, they help construct a "bridge between the province of inert matter and that of living matter, and in the place of the strict barriers previously supposed, we are warranted in presuming the existence of a multitude of gradual transitions and intermediate stages."[30]

Leduc's publication of his cumulative work in 1910 generated a renewed round of commentary. The French zoologist Lucien Cuénot, President of the Faculty of Science at the University of

Nancy, was not impressed. Although he finds the work of some interest, he notes that "the author has a manifest tendency to exaggerate the resemblances by the confusion of words" (for example, nutrition, assimilation, elimination), and he expresses some perplexity about why one would say "that the discovery [of such osmotic growths] is the discovery of the physical mechanism of the organization of living matter." "To tell the truth," he continues, "it would be difficult to understand how osmotic growths . . . could be the prelude to the formation of even very simple living beings."[31] Two years later, in a review of *La Biologie synthétique,* Cuénot's patience with Leduc is visibly strained. Now he writes, "the interpretations of M. Leduc are so fantastic . . . that it is impossible to take them seriously."[32]

For the most part, English and American reviewers were noticeably more sympathetic, even if they were a bit uncertain as to whether this was biology or not. Bashford Dean, a biologist at Columbia University, reviewed the 1910 book for *Science,* and although he begins by suggesting that "analogy" would be a more appropriate word for the title of Leduc's book than "theory," he is inclined to give the author a great deal of credit, if only for reminding us of the still unsolved problem that lies at the foundation of biology. "It may well be," he writes,

> that the post-Pasteurean biologist is over-sensitive as to the words "spontaneous generation" . . . he inclines to dismiss the rare papers which deal with the theme as anachronisms—and he is careful not to recommend them to publishers. Even the French Academy has become so modern that it will not admit to its shelves any treatise which deals with this "exploded theory"! Nevertheless, a whisper comes occasionally out of the wilderness and reminds us that this is the problem of all biological problems and that it is still neglected.[33]

The "curiously close parallels with organic processes" shown by Leduc's experiments provide just such a reminder. "Altogether," he concludes, "Leduc's book is interesting and it deserves to be carefully read. We need not admit that it *is* biology; but we must admit that the inorganic conditions which here are given detailed consideration have occurred and are occurring constantly in organisms. And we shall be apt to admit that the synthetic method promises results which will prove of great value. Leduc would be the first to agree that living substance may not be synthesized for ages, if at all. But each advance brings the goal nearer" (p. 305).

Gruenberg was less sanguine. In Part I of his two-part essay on "Artificial Life" (appearing just a few months later), Gruenberg notes that "the world was clamoring for artificial life." But as a biologist, Gruenberg holds a basic skepticism about all such efforts; he writes, "Very few of the attempts to produce 'artificial life' have been made by biologists, who realize too well the complexity of the problems involved."[34] His first discussion of Leduc's work appears in the context of this history, and a second, published two years later, in a lengthy review of *La Biologie synthétique*.[35] In both essays, Gruenberg's primary concern is to distinguish the "synthesis of 'artificial life'" from the "artificial synthesis of life," and his fear (and principal criticism of Leduc) is that Leduc has confused the two. For how, without such confusion, could Leduc possibly lay claim to having "discovered the physico-chemical foundations of life?"[36] "With wonderful patience and ingenuity Prof. Leduc has taken up in turn the commonly recognized characteristics of living cells . . . But has he thereby made an approach to the artificial synthesis of life?" Gruenberg thinks not. Nonetheless, he concludes that the work is not without merit, for "these experiments have their value in clearing the field of much conjectural rubbish and confusion."[37]

Yet another reviewer (anonymous) shows no such charity. The only entirely negative review I have been able to find in the Anglo-American press appears in *Nature,* and it is scathing.[38] Not mincing his words, the author notes, "With a little ink and water one can conjure up all sorts of phantasms . . . but," he asks, "is this sort of thing useful?" But lest this reviewer's impatience be taken as marking a difference between Leduc's American and English reception, another reviewer, in a subsequent issue of *Nature,* writes: "It is impossible to do justice to the author's arguments or make clear the proper value of his demonstrations in a short article such as the present, but this will at least serve to direct attention to a few of the very remarkable results that he claims to have achieved, which, if verified, are certainly of the highest significance to the student of the phenomena of life."[39]

The final two in this series of commentaries strike what I take to be the dominant notes. In one, entitled "The Physical Basis of Life: Laboratory Models of Living Organisms," the anonymous author begins with the question, "What is Life?" and concludes that "the field opened by Prof. Leduc is of peculiar interest and promises to bring us important disclosures regarding some fundamental life phenomena. We must in patience await further developments."[40] In the other, an accompanying editorial published in the main issue, the author locates Leduc's work in a wider controversy. He writes, "The world of biologists is divided. There are those who hold that the phenomena of life involve a separate principle which does not operate in non-living matter. Another school seeks to interpret all actions or functions of the living organism in terms of the general laws of nature which are known to apply to all matter living or dead. To us, it seems premature to take any side in this dispute." Nevertheless, he too concludes, "The line of experiment followed by Prof. Leduc is one which bears great promise of fundamental further disclosures. We have here a young branch of science, and it is the young branches which grow most actively."[41]

Clearly, Leduc's scientific colleagues, even those in the Anglo-American press, did not respond with a single voice, and the range of responses is itself of interest. But more interesting yet is the variable logic of these responses. All might agree that the similitude between his productions and living organisms is striking; they might even agree with Leduc's claim that one need only glance at the photographs of osmotic productions to recognize the forms of life. Even the geneticist William Bateson acknowledged them as "perhaps the best models of living organisms yet made."[42] But the question at issue is the scientific value, or even meaning, of such visual resemblance. What can be learned from similitude?

To one author, Leduc has offered nothing but "phantasms" which, by definition, are utterly devoid of scientific interest; to others, he has given us a biological "analogy" which nonetheless "promises results which will prove of great value" to the eventual synthesis of actual life; to still others, his "laboratory models of living organisms" promise "important disclosures regarding some fundamental life phenomena." Above all, such models promise to shed light on the as yet inscrutable chasm dividing the living from the non-living; with all the ambiguity inherent in the notion of model, they are seen as steps toward building a "bridge between the province of inert matter and that of living matter." And to others yet, they are even seen as offering "proof" that the implied chasm is itself illusory, that no such gap in fact exists, and that we are therefore "warranted in presuming the existence of a multitude of gradual transitions and intermediate stages."

Although the variations in responses to Leduc's work do not map neatly along disciplinary lines, there is a hint of such a divide in Gruenberg's observation that "very few of the attempts to produce 'artificial life' have been made by biologists." Although Leduc himself did not originally come from the physical sciences, most of his past and present colleagues in the project of synthetic biology had. Yet prior to any question about disciplinary differences in the responses to Leduc's work is the basic fact that efforts of the kind were so numerous and that they generated so much interest. Indeed, it is impossible to make sense either of Leduc's own motivation or of the interest in his work without first locating both of these in the context of ongoing debates about vitalism, mechanism, and, above all, spontaneous generation.

Spontaneous Generation in Early Twentieth-Century England

When W. Deane Butcher (past president of both the Roentgen Society and the Electro-Therapeutical Section of the Royal Society of Medicine) decided to translate Leduc's work into English, and his English and American publishers agreed to publish that translation, an interested readership was already assured. Furthermore, both Leduc and his translator had a rationale for this relatively positive response to this work in the Anglo-American scientific press. In France, they argued, Pasteur's apparent resolution of the debate over spontaneous generation still held such an iron grip on scientific politesse that a resurrection of the problem would not be tolerated in any form at all. As Butcher writes in his preface, "As recently as 1907 the Académie des Sciences excluded from its *Comptes Rendus* the report of those experimental researches on diffusion and osmoses, because it touched too closely on the burning question of spontaneous generation."[43] And at least one of the reviewers of this book (Bashford Dean) clearly concurred in this interpretation. Indeed, Leduc suggests that it is not only the issue of spontaneous generation that was a problem in France but more generally the French reception of evolutionary theory. As he wrote in his own preface, "It is undeniable that the Anglo-American people constitute a particularly favorable milieu to the birth and the development of new ideas."[44]

Leduc's instincts were good. As the reception of his work clearly indicates, he had tapped a nerve that was then very much alive in the English-speaking world. In fact, Butcher's implication that the question of spontaneous generation no longer burned in his own country may even have been a bit disingenuous. To be sure, Darwin's theory of evolution had found wide acceptance among his scientific colleagues by this time, but the majority of his English-speaking readers were nonetheless still very much preoccupied with the more particular question of spontaneous generation, even if in somewhat different terms from those in which Pasteur had posed that question. It would be more accurate to say that in the British and American context, the question had been reframed, and for some, in fact, renamed. But even with this reframing and renaming, the problem of how life began, and especially of how views about this problem related to Darwinian evolution, remained in great ferment.

By 1911 Pasteur's victory had come to hold sway not only in France but in British and American scientific circles as well. To be sure, British science could boast one prominent hold-out, namely, the renowned microbiologist Henry Charlton Bastian. Famous (or infamous) from his debate with Pasteur in the 1870s, Bastian had never given up, and he continued in his efforts to demonstrate the spontaneous generation of bacteria and other microorganisms until his death in 1915. But in the last decades of his life, he was effectively alone in his advocacy.[45] Indeed, Bastian's isolation had begun long before. James Strick argues that Bastian's marginalization—together with the effective severance of spontaneous generation from Darwin's theory *tout de suite*—came as a direct consequence of the campaign by Thomas Huxley and John Tyndall in the 1870s to improve "the public profile of the Darwinian camp."[46] The influence of Huxley and Tyndall was enormous, and it is perhaps a measure of that influence that, as Strick also observes, "it has been widely supposed for more than a century now that no respectable supporter of Darwin seriously advocated spontaneous generation in mid-Victorian England." In fact, however, even in the nineteenth century, "many evolutionists in Britain . . . agreed with Bastian that spontaneous generation and evolution were linked theories."[47] By the early part of the twentieth century that linkage had become even more acceptable. Ironically, it was Huxley himself who, in his 1870 presidential address to the British Association for the Advancement of Science (BAAS), provided both a rhetorical and a conceptual opening for the reestablishment of such a linkage: first, by distinguishing the occurrence of spontaneous generation in the past from its occurrence in the present and, second, by introducing a new term to refer to "the hypothesis that living matter may be produced by not living matter," namely, abiogenesis, and reframing the debate as one between biogenesis ("the hypothesis that living matter always arises by the agency of pre-existing living matter") and abiogenesis. In his terms, spontaneous generation referred only to the older hypothesis, disproven by Pasteur, that living matter *is now* arising from non-living matter.[48]

A good summary of views dominant in the early part of the twentieth century can be found in the 1911 edition of the *Encyclopedia Britannica,* where, under the term "spontaneous generation," the index directs the reader to the heading "abiogenesis." The author of the entry, P. Chalmers Mitchell, explains that "abiogenesis," the modern equivalent of "spontaneous generation," is "the term . . . for the theory according to which fully formed living organisms sometimes arise from not-living matter." And he concludes,

> So far, the theory of abiogenesis may be taken to be disproved. It must be noted, however, that this disproof refers only to known existing organisms . . . It may be that in the progress of science it may yet become possible to construct living protoplasm from non-living material. The refutation of abiogenesis has no further bearing on this possibility than to make it probable that if protoplasm ultimately be formed in the laboratory, it will be by a series of stages, the earlier steps being the formation of some substance, or sub-

> stances, now unknown, which are not protoplasm. Such intermediate stages may have existed in the past.[49]

The question at issue was therefore less one of the possibility of living beings (as we now know them) emerging directly and fully formed from non-living matter, and more one of the spontaneous emergence, even now, of some sort of intermediate forms—forms that might shed light on how life had originated in the past. But notice Mitchell's interesting use of the terms proof, disproof, and theory and his easy slide, just as in the remarks of Loeb and Leduc, from an understanding of the origin of life to the actual construction of life in the laboratory. Here, however, we might gain some insight into the reasoning behind such a move. As Mitchell presents it, the theory of abiogenesis is a claim not of theoretical possibility but of actuality—not of what might but of what in fact does, at least sometimes, occur. Hence, the conjunction between proof (or explanation) and construction: to an experimental biologist, such a theory can only be proven by the demonstration of the occurrence of the actual phenomenon in the laboratory. Given that demonstrations of the emergence of fully formed organisms have ("so far") failed, the question that remains for Mitchell is whether more rudimentary quasi-forms of life (or proto-organisms) can be shown to emerge spontaneously. For, as he discreetly puts it, "such intermediate stages may have existed in the past." His discretion is noteworthy.

Even in England (Darwin notwithstanding), the question of whether life had originally emerged spontaneously (be it by stages or not) remained far from easy, arousing quite vigorous and often conspicuously charged debate. An indication of just how lively such debates often were can be gleaned from both the scientific and lay responses to Sir Edward Albert Schäfer's 1912 presidential address to the BAAS. The subject of Schäfer's address was nothing less than "the nature, origin, and maintenance of life."[50] A renowned physiologist and an eminently qualified representative of the scientific biology of his time, he begins with the standard materialist credo:

> The problems of life are essentially the problems of matter; we cannot conceive of life in the scientific sense as existing apart from matter. Living phenomena are investigated, and can only be investigated, by the same methods as all other phenomena of matter, and the general results of such investigations tend to show that living beings are governed by laws identical with those which govern inanimate matter. The more we study the manifestations of life the more we become convinced of the truth of this statement and the less we are disposed to call in the aid of a special and unknown form of energy to explain these manifestations.

Yet the assertion of a physical basis of vital phenomena remained a far cry from the denial of any discontinuity at all between the living and the non-living. Many of his colleagues would have fully embraced the former while continuing to balk at the latter. And indeed, it was precisely his claim that, just as in the past, so too in the present, life could evolve from non-life in a smooth transition, through stages "which we shall be uncertain whether to call animate or inanimate," that so aroused the British press, eliciting such epithets as "revolutionary" and "epoch-making."[51]

Furthermore, for Schäfer just as for Mitchell, the only available and hence the only legitimately scientific test of such a claim (or theory) lay in the possibility of constructing transitional forms in the laboratory.[52] Because a search for the traces of such forms in the geological record seemed so futile ("We can only expect to be confronted with a blank wall of silence"), the very possibility of a scientific solution of the problem of the origin of life required the assumption that the process was not a unique event but could be repeated, either naturally or synthetically. "If living matter has been evolved from lifeless in the past, we are justified in accepting the conclusion that its evolution is possible in the present and in the future. Indeed, we are not only justified in accepting this conclusion, we are forced to accept it. When and where such change from non-living to living matter may first have occurred, . . . when or where it may still be occurring, are problems as difficult as they are interesting, but we have no right to assume they are insoluble." And because we "are not likely to obtain direct evidence regarding such a transformation of non-living into living matter in Nature,"[53] the sole remaining option and, indeed, the primary challenge (even obligation) for scientists, is therefore an experimental one, albeit one guided by "the mind's eye": evidence must be sought in the production, under laboratory conditions, of the missing transitional forms, perhaps especially of those forms "which we shall be uncertain whether to call animate or inanimate."

Crossing the Channel

Leduc could scarcely have hoped for a better introduction. *The Mechanism of Life* had appeared only the year before, and indeed it was twice cited by Schäfer in support of his claims. Even so, translation from Leduc's concerns to those of his biological colleagues on the other side of the channel was not without its pitfalls, for differences of moral and epistemological imperatives, as well as of intellectual history, all had to be negotiated. The most immediately apparent of these is the difference in intellectual history which was itself a product of national difference.

Like Schäfer and Mitchell, Leduc's focus was on the question of intermediate forms. And unlike Bastian, his aim was not to prove Pasteur wrong but rather to insist on the question that had been left hanging by Pasteur's apparent victory. How, if not by spontaneous generation, could living beings ever have originated in the first place? For Leduc, as for Schäfer and his colleagues, the goals of a scientific biology required that this question be addressed in the context of evolutionary theory. But Leduc's evolutionary theory was not quite the same as Schäfer's and Mitchell's. As Leduc saw it, the doctrine of evolution originated with J. B. Lamarck, and Darwin's subsequent contributions were scarcely more than elaboration of the original theory. From such a perspective, locating the origin of life in an evolutionary perspective accordingly meant locating it in the evolutionary perspective of Lamarck. Moreover, given how closely linked the issues of spontaneous generation and evolution were in Lamarck's own writings, to rehabilitate the former was to rehabilitate the latter, and the rehabilitation of both was synonymous with the rehabilitation of Lamarck himself. Unfortunately, he wrote, in France the doctrine of evolution was smothered "under the weight of authority of [Lamarck's] adversaries." "Before [it] could live and take its proper place, it had to be reborn in England—the country of liberty. This resuscitation was due to Darwin." Making the point yet more explicitly, Leduc writes: "Lamarck's work was still-born, whereas that of Darwin lived and grew to its full development. This was due, not to any imperfection or insufficiency in Lamarck's work, but to the milieu into which it was born."[54]

Leduc, like Schäfer, makes clear his commitment to the physical-chemical basis of life in his introductory remarks: "Living things are made of the same chemical elements as minerals; a living being is the arena of the same physical forces as those which affect the inorganic world."[55] In these same remarks, he also articulates his further commitment to "the doctrine of Evolution" as he interprets that doctrine: "The chain of life is of necessity a continuous one, from the mineral at one end to the most complicated organism at the other. We cannot allow that it is broken at any point, or that there is a link missing between animate and inanimate nature. Hence, the theory of evolution necessarily admits the physico-chemical nature of life and the fact of spontaneous generation. Only thus can the evolutionary theory become a rational one."[56]

But it is not until his concluding remarks that Leduc spells out his commitment to Lamarck as "the true originator of the scientific doctrine of evolution" (p. 160), and with that commitment, his reliance on Lamarck's writings for this particular understanding of evolutionary theory. Here he writes, "Without the idea of spontaneous generation and a physical theory of life, the doctrine of evolution is a mutilated hypothesis without unity or cohesion. On this point Lamarck speaks most clearly: '. . . Nature herself possesses all the faculties and all the means of producing living beings in any variety. She is able to vary, very slowly but without cessation, all the different races and all the different forms of life, and to maintain the general order which we see in her works.'" Based on his reading of Lamarck, Leduc now reiterates his opening dictum: "The doctrine of evolution should reconstitute every link in the chain of beings from the simplest to the most complicated; it cannot afford to leave out the most important of all, viz. the missing link between the inorganic and the organic kingdoms. If there is a chain, it must be continuous in all its parts, there can be no solution of continuity" (p. 165).

Evolutionists like Lamarck and Haeckel admit spontaneous generation, not as the most probable but as the only possible explanation of the phenomenon of life: "Lamarck shows us the apparition of living things at a certain epoch of the earth's evolution, and the gradual development of more complicated form . . . Darwin shows how heredity and natural selection tend to accentuate the variations which are favourable to existence . . . These are great and admirable conquests of the human intelligence, they have demonstrated the first appearance and the progressive evolution of living beings; it now only remains for us to explain them" (pp. 165–166). Yet interwoven with Leduc's epistemological imperative is also a moral imperative, clearly signaled by his explicitly moral injunctions (for example, we "cannot afford," we "cannot allow"). Schäfer's remarks ("we have no right") similarly indicate a moral imperative. But the nature of that imperative is noticeably different in the two texts.

For Schäfer, what "forces" us to accept the conclusion that the evolution of the living from the non-living remains possible in the present and in the future, that which denies us the "right to assume [the problems of the origin of life] are insoluble," is nothing other than the moral imperative of science. To quote an accompanying editorial in the *Scientific American Supplement,* "The position taken by Prof. Schäfer . . . is, that until all the resources of the methods of investigation of the physical sciences are exhausted, we have no occasion to refer to any 'unexplained' effect to supra-physical causes."[57] Leduc's moral imperative comes from elsewhere. His claim that the doctrine of evolution "cannot afford" any gaps, that "we cannot allow that [the chain of life] is broken at any point, or that there is a link missing between animate and inanimate nature," that "if there is a chain, it must be continuous in all its parts" derives not from a methodological commitment but from at least equally powerful ontological and theoretical commitments. Indeed, like many physicists of his own time and later, Leduc scarcely distinguishes between the two: natural law and natural history are one and the same. Here, his commitments to evolutionary theory, to Lamarck, and to the continuity of nature all merge into a single necessity. And from such a fusion, it follows "naturally" that what the doctrine of evolution "cannot afford," we too "cannot allow."

The last difference between Leduc and Schäfer to which I call attention bears directly on the question of what ought to count as experimental evidence for the continuous evolution of living from non-living. For Schäfer, the primary contribution of Leduc's experimental efforts is conceptual rather than evidential. Although "everybody knows, or thinks he knows, what life is,"[58] Leduc's osmotic phenomena show us just how difficult such a categorical distinction is; above all, they help subvert the assumption that the properties most commonly taken as characteristic of life are distinctively vital. But nowhere does Schäfer suggest that Leduc's productions might attest to the process by which life on earth actually evolved. For Schäfer, such evidence would have to come from the observation, under laboratory conditions (just because natural conditions precluded such observation), of forms that were literally and materially at least proximal to life as we know it. That is to say, the laboratory productions would have to reproduce not just the formal properties of living beings but also the same material properties that had already been identified. For Leduc, by contrast, formal (or visual) proximity could by itself bear witness to the evolutionary process that had actually occurred.

Denouement

Even after the initial flurry of popular interest had passed, Leduc's work continued to be cited in more specialized books and articles for another fifteen or more years. William Bateson, for example, cited Leduc's osmotic growths in his Silliman lectures on *Problems of Genetics* as "perhaps the best models of living organisms yet made."[59] The work also figured prominently in the 1918 book by the botanist John Muirhead Macfarlane, *The Causes and Course of Organic Evolution*. Here, Macfarlane seconds Leduc's claim that his structures are "fully analogous to that which we meet in a living organism" and takes strong exception to critics like Le Bon who see "'hardly any more connection between these artificial plants and the real ones than there is between a living man and his statue.' We would rather suggest," writes Macfarlane, "and we hope to show, that these are the necessary stages and phenomena that carry us from inorganic colloids . . . to the varied series of organic colloids in which . . . biotic energy resides."[60]

Similarly, Leduc's efforts were frequently cited by the physiologist R. S. Lillie in a series of articles on his own researches into the physical-chemical mechanisms of organic growth.[61] For Lillie, brother of F. R. Lillie and a far more mainstream figure in the American biological community than Macfarlane, the central point is obvious: "The study of the structure and properties of growing inorganic systems . . . may thus be expected to throw some light upon the more general features of the growth-process in organisms. Such systems may be regarded as elementary or generalized models of organic growth."[62] Leduc's last published work on the subject, "Solutions and Life," appeared in 1928, back to back with an essay by Herrera on "Plasmogeny," in a collection of articles on *Colloid Chemistry* edited by the eminent chemist Jerome Alexander.[63] And as late as 1938, R. Buettner, a professor of pharmacology at the Hospital of Philadelphia, included an extensive discussion of Leduc's work in his book, *Life's Beginning on the Earth.* This he concludes as follows: "In spite of many striking life-like features, none of these artificial structures is a living entity. They all lack the power of propagation. Moreover they are too brittle to maintain themselves for any considerable time. And yet we may learn a great deal from these perishable artificial structures. In a striking manner they reveal the widespread action of the formative forces which nature has at its command . . . The secret of life is thus unfolded a little."[64]

But Buettner's relatively favorable notice was something of a last gasp for Leduc's synthetic biology. The judgment that had come to prevail by the late 1930s was far closer to that of Aleksandr Oparin. Because of his own work on the origin of life, because of his commitment to the gradual emergence of life out of the ever more complex chemical dynamics of metabolism, and because of his particular belief that life evolved out of pre-vital colloidal structures, Oparin might have been expected to show interest in Leduc's work, but his support was not forthcoming.[65] Indeed, in his *Origin of Life* (published in Russian in 1936 and translated into English in 1938), Oparin cites Leduc's work only to dismiss it at once: "As an illustration of such an excessive delusion by external appearances, Stéphane Leduc's experimental productions of so-called osmotic cells may be mentioned." Unlike Macfarlane, he endorses the sentiments of Le Bon (and even invokes Le Bon's very image) as he goes on to note that the resemblance of these productions to living cells is "not greater than the external resemblance between a live person and his marble image."[66] Judging by the citation record, history concurred with Oparin. And most contemporary readers, were they to know about Leduc's work at all, would almost certainly concur as well.

Yet the question of whether a resemblance is great or small, delusory or instructive, compelling or inconsequential, or a matter for laughter or tears, is hardly so simple. Leduc's artificial organisms were certainly (in fact, by definition) not real organisms, neither to him nor to his readers. They were analogies, models, simulations. To be sure, a model is expected to bear some resemblance to that which is being modeled, but in science as in art, the degree of resemblance is generally understood to be a matter of perspective. The more critical question is whether it is a "good" model, and in both science and art the measure of how good a model is varies notoriously.

If it is possible to make any generalizations at all, one might say that it is here, in the criteria brought to the measure of "goodness," that models in science and art most clearly depart. For the value of a scientific model is judged, first and foremost, by its utility. Thus, the issue is rarely one of any absolute degree of resemblance (however that might be measured) but whether or not what resemblance it does bear is close enough in ways that make it possible for the model to be scientifically productive. There is a lot of hedging here, but—once one recognizes the enormous variability in the meaning of scientifically productive—necessarily so.

No one doubted that Leduc's osmotic growths bore a close resemblance to biological organisms: the question at issue was whether or not the resemblance was meaningful, whether it was close enough and of the right kind to permit observations in the domain of the model to qualify as answers to questions about the nature and origin of real life which were then of scientific interest. During the peak of Leduc's publications, opinion was clearly divided. But writing at the end of the 1930s, Oparin's response proved to be definitive, for by that time the voices that could be heard in Leduc's defense were few indeed. Yet not many years before, a quite substantial number of respected scientists (including biologists) thought the resemblance between Leduc's creations and actual living beings was sufficiently strong to be compelling, illuminating, instructive, and even explanatory. What changed?

For one thing, the questions changed. In the early part of the twentieth century, discussions about the origin of life, and even about spontaneous generation, were compelling precisely because these issues bore so directly on what was then the question of all biological questions: Can the phenomenon of life be fully explained by physical and chemical forces, or does it require a separate principle, one not operating in non-living matter? In 1913, as the editorial in *Scientific American* baldly put it, "The world of biologists [was] divided." Despite numerous successes in identifying the physical and chemical basis of vital phenomena, the processes that were then seen as most basic to organismic life—growth and reproduction, and perhaps especially morphogenesis and development—appeared totally resistant to physicalist explanation. Faced with this question, Leduc's efforts were far from silly: while they may not have been useful in identifying the actual physical basis of such phenomena, at the very least they demonstrated that purely physical dynamics could, in principle, give rise to processes so similar as to void the most obvious complaint.

Twenty-five years later, however, the battle against vitalism had been won, and if some biologists remained convinced of an ineradicable divide between life and non-life, for the most part they kept such convictions to themselves. To be sure, the problems that Leduc had sought to confront (such as morphogenesis, the origin of life) could scarcely be said to have been solved. But with the threat of vitalism laid to rest, they had clearly lost their urgency. Furthermore, other questions—especially questions about the mechanism of heredity—had come to take center stage, and to these questions Leduc's method could contribute nothing at all.

One last point, and it is to note the frequency with which such descriptors as marvelous, wonderful, spectacular, strange, and amazing were employed by Leduc and his readers. Indeed, even Leduc's detractors' insistence that his creations bore no greater resemblance to actual organisms than that between "a live person and his marble image" summons forth, merely by the invocation of entrenched myths and literary motifs in which statues do come to life (or in which living beings are turned to stone), the very sense of wonder these critics wished to squelch. One is reminded of Freud's famous essay on "The 'Uncanny.'"[67] Reviewing "the things, persons, impressions, events and situations which are able to arouse in us a feeling of the uncanny in a particularly forcible and definite form," Freud, citing an earlier essay by E. Jentsch (1906), writes: "Jentsch has taken as a very good instance 'doubts whether an apparently animate being is really alive; or conversely, whether a lifeless object might not in fact be animate'; and he refers in this connection to the impression made by wax-work figures, ingeniously constructed dolls and automata" (p. 226). Jentsch had also observed, correctly from

Freud's perspective, that a particularly successful narrative device for "creating uncanny effects is to leave the reader in uncertainty whether a particular figure in the story is a human being or an automaton, and do it in such a way that his attention is not focused directly upon his uncertainty" (p. 227). But was it not precisely such uncertainty, the identification of forms or stages "which we shall be uncertain whether to call animate or inanimate," that Edward Albert Schäfer had argued was required to close the gap between the living and the non-living? Neither Leduc nor the readers who found his work "of the highest significance" to the study of life claimed that his osmotic growths were alive; yet what made the resemblances so compelling, what made them wonderful, marvelous, amazing, may have been just such residual uncertainty as to their actual status. So close did these imitations seem that they could even, as Butcher wrote, "deceive the very elect." Yet the potential for deceit claimed here was meant as a positive virtue, not a negative one: it made possible that "willing suspension of disbelief" that permits uncertainty to remain out of focus, that allows the "as if" to do the remarkable work it has so often done in the past, and perhaps especially in the growth of the physical sciences.

As James Clark Maxwell had once written in his reflections upon the fictional status of the mathematical function representing the electric potential, "We have no reason to believe that anything answering to this function has a physical existence in the various parts of space, *but* it contributes not a little to the clearness of our conceptions to direct our attention to the potential function *as if* it were a real property of the space in which it exists."[68] Leduc's models were material rather than mathematical, but the point, I think, is the same.

CHAPTER TWO

Morphology as a Science of Mechanical Forces

[Natural] Philosophy is written in this grand book, the universe, which stands continually open to our gaze. But the book cannot be understood unless one first learns to comprehend the language and read the letters in which it is composed. It is written in the language of mathematics, and its characters are triangles, circles, and other geometric figures without which it is humanly impossible to understand a single word of it; without these, one wanders about in a dark labyrinth.

Galileo, "The Assayer" (1623)

The study of Form may be descriptive merely, or it may become analytical. We begin by describing the shape of an object in the simple words of common speech: we end by defining it in the precise language of mathematics; and the one method tends to follow the other in strict scientific order and historical continuity.

D'Arcy Wentworth Thompson, "Morphology and Mathematics" (1915)

o one would deny the centrality of an understanding of the nature and origin of form to anything purporting to be an explanation of biological development. But the question of what ought to qualify as a scientific study of form has been among the most fraught of twentieth-century biology. Can the study of organic form ("morphology," as Goethe named it) ever become a "science of form"? And if so, what would it take? More specifically, what is the proper role of physical and mathematical models in the development of such a science? In the passage cited above, D'Arcy Wentworth Thompson followed Galileo, arguing that the study of form would become a science only when it was expressed in the

language of mathematics; elsewhere, he invoked another (and to him equivalent) criterion of a true science of form, namely its description in terms of physical mechanisms.[1] Stéphane Leduc's vision of scientific progress made no reference to mathematics, but he and Thompson found common ground in the belief that an understanding of biological form requires understanding its physical mechanisms. For both men, the use of models and metaphors drawn from physics and chemistry were essential.

That Leduc's models for the growth and development of living organisms fared so poorly will come as no surprise to today's student of biological development, and not just because of the conspicuous artificiality of his osmotic growths. At least equally important is the fact that, in contradistinction with the physical sciences, material models, be they mechanical or chemical, have played so slight a role in the history of developmental biology. And despite their prominent use in population biology, mathematical models have played an even smaller role.[2] Indeed, the foremost meaning contemporary developmental biologists are likely to associate with the term model is neither a mechanical or chemical model nor a set of equations: it is an organism. Model organisms (such as the fruit fly, *Drosophila melanogaster;* the round worm, *Caenhorabditus elegans;* and the house mouse, *Mus musculus*) are an explicit preoccupation of a great deal of literature today, both scientific and historical.[3] Furthermore, as anyone familiar with the history of research in twentieth-century life sciences knows, the use of exemplary organisms to represent a class of biological phenomena (whether genetical, developmental, or behavioral) across a wide range of species is hardly new. But unlike mechanical and mathematical models (and this may be the crucial point), model organisms are exemplars or *natural* models—not artifactually constructed but selected from nature's very own workshop.[4]

Other differences are also worth noting. The primary criterion for the selection of a model organism is only rarely its simplicity—the principal criterion for a model in the physical sciences. Far more important is the experimental accessibility endowed by particular features (such as size, visibility, reproductive rate). Furthermore, its primary use is neither for the construction of a general theory nor as a guide to identifying the leading causal factors for a particular process. Model organisms represent in an entirely different sense of the word than do models in the physical sciences: they stand not for a class of phenomena but for a class of organisms. As such, they are more closely akin to political representatives, and, in fact, are employed in a similar fashion—as a way of inferring the properties (or behavior) of other organisms. It is for just this reason that biological modeling has sometimes been described as proceeding "by homology" rather than "by analogy."[5]

The success of D'Arcy Wentworth Thompson's magnum opus, *On Growth and Form,* presents something of a puzzle for historians of science. Thompson was a zoologist and comparative anatomist—he trained at Cambridge University with Michael Foster and Francis M. Balfour, served as chair of the Biology Department at the University of Dundee from 1884 to 1917 and as chair of Natural History at the University of St. Andrews from 1917 to his death in 1948, and was knighted in 1937. By avocation, he was a classicist. But it is undoubtedly for his book *On Growth and Form,* first published five years after Leduc's *La Biologie synthétique,* that he is best remembered today. Like Leduc, Thompson sought an understanding of biological processes through their similitude with mechanical and chemical processes, and he made extensive use of many of Leduc's own photographs. But unlike Leduc, Thompson has had a long and illustrious place in the annals of twentieth-century science. What is the basis of his enduring renown? How are we to account for the high regard in which his celebration of the importance of physical and mathematical models in biology has been held, when the actual use of such models in the development of modern biology has in fact been so minimal?

More specifically, how are we to account for the contrast between Thompson's success and Leduc's manifest failure? A number of differences between these two authors are readily identifiable. Thompson was a biologist by training with a prominent position in British academia, and he was an elegant essayist and master of a vast sweep of biological lore. Moreover, his long argument was primarily a synthesis of the work of others, with little suggestion of promoting his own scientific accomplishments. By contrast, Leduc was a physician by training, working on the margins of biology; his primary intellectual context was French, and one hostile to his efforts at that. In fact, his turn to an English-speaking audience was in the first instance motivated by that hostility. Finally, his writing not infrequently invited the charge of tendentious self-promotion. All of these differences surely worked against Leduc and in Thompson's favor.

But other differences, less obviously favorable to Thompson, can also be seen. Leduc's focus was manifestly both experimental and synthetic, devoted to actual laboratory productions, while Thompson's concerns tended to be more theoretical and even philosophical. For example, there is no evidence that Thompson had any interest whatsoever in synthesizing life. Where Leduc sought an experimental procedure that would enable man, "by directing the physical forces which are its cause," to reproduce the phenomenon of the origin of life,[6] Thompson's aims were infused by more abstract concerns—conceptual and aesthetic rather than instrumental. The representation of living processes in mathematical form might have utilitarian value, but it could also be viewed as an end in itself. Similarly, while the two men shared an interest in morphogenesis, along with the hope of explaining this quintessentially biological phenomenon by the action of physical forces, the very concept of form had different resonances for the two authors. For Leduc, the primary meaning of form was physical construction, as in "to form." For Thompson, the word also carried other, more classical, meanings; along with physical construction, it evoked as well both the aesthetic notion of "formal" and the mathematical notion of "formalism." Thompson's approach to the study of biological form was influenced not only by the success of mathematics in the physical sciences; it was stamped also by the traditions of his own discipline, morphology, as well as by the traditions of the Greek classics that were his lifelong avocation (perhaps especially, "by the dreams and visions of Plato and Pythagoras").[7] It is precisely this multiplicity of concerns that has lent Thompson his appeal to such a diversity of readers: his interest in construction arouses the sympathy of bioengineers; his interest in the aesthetics of form attracts the attention of artists and architects; and his interest in mathematical formalism invites contemporary mathematical biologists to claim him as the forefather of their discipline.

What did Thompson's work actually consist of? As we take a look at his arguments in some detail, we can begin to see the variety of ways in which those arguments might have been construed as contributing to an understanding of growth and form.

On Growth and Form: An Overview

The first key to Thompson's agenda is indicated in the very title of his work. The study of form cannot be separated from the study of growth. "How a thing grows, and what it grows into," he wrote, "is a dynamic and not a merely material problem; so far as the material substance is concerned, it is only by reason of the chemical, electrical or other forces which are associated with it."[8] The form of an organism cannot be understood merely in terms of its material substructures; rather, it requires an understanding of the forces that operate between these substructures. And for this, the experience of the physical sciences has abundantly demonstrated the value of mathematics. By his own summary description, his aim was thus "to study the inter-relations of growth and form, and the part which the physical forces play in this complex interaction; and, as part of the same enquiry, to use mathematical methods and mathematical terminology to describe and define the forms of organisms." Furthermore, while his primary aim was "to shew the naturalist how a few mathematical concepts and dynamical principles may help and guide him," he hoped as well "to shew the mathematician a field for his labour—a field which few have entered and no man has explored."[9]

But behind this apparently unproblematic agenda lay a mission, and it was with something of a missionary's zeal that he set out to counter the widespread skepticism of his colleagues, even to heal the "parting of the ways" which "the introduction of mathematical concepts into natural science has seemed to so many men" (p. 11). As he wrote: "The zoologist or morphologist has been slow . . . to invoke the aid of the physical or mathematical sciences . . . To treat the living body as a mechanism was repugnant, and seemed even ridiculous . . . [The zoologist] would fain refer to psychical forces . . . In short, he is deeply reluctant to compare the living with the dead, or to explain by geometry or by mechanics the things which have their part in the mystery of life" (pp. 2–3). Joseph Needham has suggested that "it was precisely because D'Arcy Thompson considered morphology the last stronghold of irrational views that he devoted his life to the mathematization of it."[10]

Yet Thompson was not willing to go quite so far as Leduc in his repudiation of extra-physical forces in the phenomena of life, and at least initially this may well have been one factor in his success. He was careful to allow the organism a space, small to be sure, but still large enough, as Niels Bohr once put it, "to permit [the organism], so to say, to hide its ultimate secrets from us"[11]—and to permit readers to hide their own lingering beliefs in vital forces. Thompson wrote, "I would not for the world be thought to believe that this is the only story which Life and her Children have to tell."[12] In other words, it was not "life itself" but its workings that he sought to illuminate, not life's origin but its mechanisms of growth and form. However living beings might have originated, and whatever other principles might ultimately need to be brought to bear, biological organisms were in the first instance constituted of matter. Yet even here, he remained ecumenical. Analysis of efficient causation might or might not be sufficient to account for living phenomena in general; he argued only that it was necessary. For phenomena such as change in form, it might even be adequate.

Thus, the mission he took upon himself was not to demonstrate the absence of *any* distinction between the living and the non-living but merely to show that the physical and mathematical sciences provided the necessary starting point for the understanding of the structure and dynamics of organic form: "Of the construction and growth and working of the body, as of all else that is of the earth earthy, physical science is, in my humble opinion, our only teacher and guide" (p. 13). And a little later, "The form, then, of any portion of matter, whether it be living or dead, and the changes of form which are apparent in its movements and in its growth, may in all cases alike be described as due to the action of force. In short, the form of an object is a 'diagram of forces'" (p. 16).

The question is how to demonstrate the adequacy of mechanical forces to account for the growth and form of biological organisms. He could not proceed from the physical properties of living material for the simple reason that adequate measurements were not available. Accordingly, Thompson (like Leduc) relied heavily on the persuasive powers of visual analogy. Observing the striking resemblances he presents between, for example, the phases of a splash of liquid and the shape of a hydroid polyp; the shapes produced by a falling drop of fusel oil in paraffin and the forms of medusoids; the shape of a cube of gelatin upon drying and that of a shark vertebra; the structure of soap bubbles and the patterns of a segmenting egg, one can scarcely avoid being impressed.[13] Like Leduc's osmotic growths (which Thompson cites on numerous occasions), the visual similarity of these arti- factual productions to the natural products of biology makes a case for, at the very least, the possibility of establishing living forms of nature as "diagrams of forces." As Robert Olby has emphasized, the crucial point for Thompson was that such structures were not predicated on subcellular, pre-formed structures but were simply the outcomes of familiar physical processes.[14] Where it was possible to establish at least some degree of correspondence between the physical conditions underlying the two kinds of structures, the force of the visual analogy could be strengthened even further. Indeed, so effective are some of his examples that readers may find themselves inexorably drawn to a more certain conviction—not merely of possibility but of actuality. Invoking Thompson's own paraphrase of Bacon, one could say that such examples do not merely allure, rather they extort assent.[15]

His considerations of mechanical efficiency in relation to bone structure provide a case in point, and his report of an encounter between a structural engineer and an anatomist so lucidly illustrates the force of his diagrammatic juxtaposition of the structures of a crane-head and femur (see Figure 3) that I quote it here almost in its entirety:

> A great engineer, Professor Culmann of Zurich . . . happened (in the year 1866) to come into his colleague Meyer's dissecting room, where the anatomist was contemplating the section of a bone. The engineer, who had been busy designing a new and powerful crane, saw in a moment that the arrangement of the bony trabeculae was nothing more nor less than a diagram of the lines of stress . . . in the

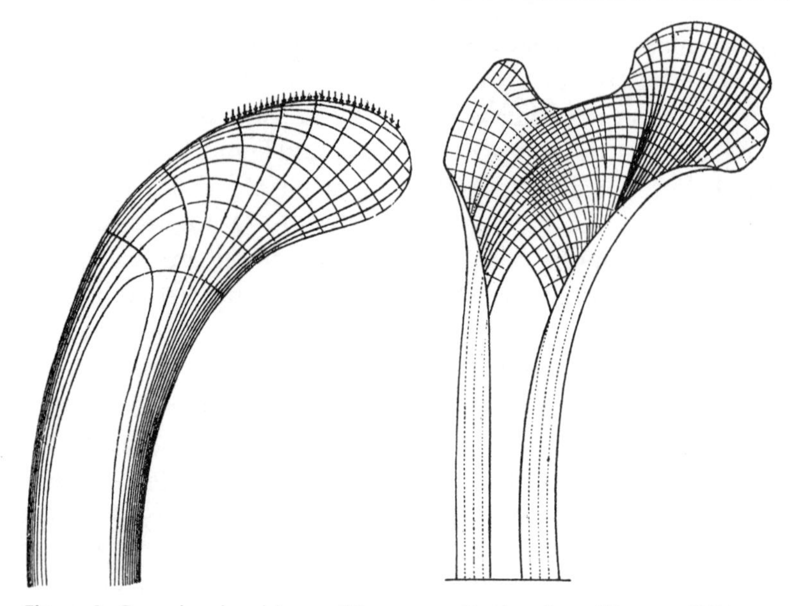

Figure 3. Crane-head and femur. (Thompson, *On Growth and Form*, p. 978.)

> loaded structure: in short, that Nature was strengthening the bone in precisely the manner and direction in which strength was required; and he is said to have cried out, "That's my crane!" (pp. 976–977)

Thompson of course recognizes that the comparison is not quite that straightforward: "The head of the femur is a little more complicated in form and a little less symmetrical than Culmann's diagrammatic crane." Nevertheless, the force of the analogy is clear enough. As he goes on to explain, "we have no difficulty in seeing that the anatomical arrangement of the trabeculae follows precisely the mechanical distribution of compressive and tensile stress or, in other words, accords perfectly with the theoretical stress-diagram of the crane" (p. 978).

Diagrams of force (or stress) thus provide a window into the physical processes by which bodies are formed—a mode of analysis heretofore absent from morphology but essential to its proper understanding. They give us access not only to the process by which individual parts of a body come into being but even more importantly to the design of the overall structure of an organism. The structure of an organism is a function of the particular physical conditions in which it lives; and unless one attends to the constraints these conditions impose on the skeleton as an integrated structure, one is bound to be led astray. "The whole skeleton and every part thereof, down to the minute intrinsic structure of the bones themselves, is related in form and in position to the lines of force, to the resistances it has to encounter." The various parts that make up the whole fabric of the body "are moulded one with another; they come into being together, and act and react together. We may study them apart, but it is as a concession to our weakness and to the narrow outlook of our minds . . . There can be no change in the one that is not correlated with changes in the other . . . They are *parts* of a whole which, when it loses its composite integrity, ceases to exist" (pp. 1018–1019). Without an analysis of the physical constraints on bodily structure, no understanding of phylogenetic relationships can be complete. The importance of heredity to the science of morphology can not be denied, but "to see in the characters of a bone merely the results of variation and of heredity" is to risk "error and misconception"; it is "to give to that science a one-sided and fallacious simplicity" (pp. 1022–1023).

Curiously, however, Thompson's final argument against the sufficiency of variation and heredity for an understanding of phylogenetic relationships, and especially against Darwin's view of evolution as an accumulation of small variations, has little to do with physics or engineering. This is his "theory of transformations," and it represents a different sort of analysis altogether. Relying on geometry rather than on force diagrams, Thompson here most nearly approaches his ideal of a mathematical formulation that can represent the fundamental simplicity underlying the form of material things.[16] Despite the fact that it appears only in his concluding chapter, the theory of transformations is at once the contribution with which Thompson's name is most frequently associated and that which draws most directly on his own work. The gist of his theory is an attempt to show that families of animal forms are bound by an underlying algebraic structure: within a family, the morphology of one animal could be mapped onto that of another by a simple transformation of coordinates. Yet there are limits to such mapping. No coordinate transformation will map the form of an organism belonging to one family onto that belonging to another:

> We cannot fit both beetle and cuttlefish into the same framework, however we distort it; nor by any coordinate transformation can we turn either of them into one another or into the vertebrate type. They are essentially different; there is nothing about them which can legitimately be compared. Eyes they all have, and mouth and jaws; but what we call by these names are no longer in the same order or relative position; they are no longer the same thing, there is no *invariant* basis for transformation. (p. 1086)

Between distinct families thus seems to lie a fundamental discontinuity that "weigh[s] heavily against Darwin's conception of endless small continuous variations." Such "geometric analogies," he concludes, "help to show that discontinuous variations are a natural thing, that 'mutations'—or sudden changes, greater or less—are bound to have taken place, and new 'types' to have arisen, now and then" (pp. 1094–1095).

D'Arcy Thompson's Reception

D'Arcy Thompson originally wrote *On Growth and Form* in the midst of the First World War; during World War II, he revised and expanded the original edition (from 793 to 1116 pages),[17] and the revised edition has been reissued at regular intervals ever since. From the beginning, its contribution was clearly recognized as more synthetic than original, leaning heavily on the work of others, but that recognition has done little to dampen the ardor of either his early or more recent fans. From the time of its first publication in 1917, the praise elicited by *On Growth and Form* has been both profuse and extravagant—one might say, hyperbolic. The prominent zoologist J. Arthur Thomson was particularly enthusiastic. In one of the first reviews to appear, he wrote, "This book, at once substantial and stately, is to the credit of British science and an achievement for its distinguished author to be proud of . . . We offer Prof. D'Arcy Thompson felicitations on his masterly book. It marks a big advance in science, and it will make other advances possible . . . His argument, couched in a style that is always clear and dignified, and at times, bewitchingly beautiful, has given us a fresh revelation of the unity of Nature."[18]

P. B. Medawar's response to the second edition was equally enthusiastic. He described the work as "beyond comparison the finest work of literature in all the annals of science that have been recorded in the English tongue."[19] And in a similar vein, G. Evelyn Hutchison once referred to it as "one of the very few books on a scientific matter written in this century which will . . . last as long as our too fragile culture."[20]

Hutchison might even have been right. On the fiftieth anniversary of D'Arcy Thompson's death in 1998, two international conferences were hosted to celebrate his life and work (one on "The Mathematical Biology of Pattern and Process" in Bath, in April, and the other, on "Spatio-temporal Patterning in Biology" in Dundee, in September), and for many scientists his name endures as something of a touchstone for the very idea of mathematical biology.

Yet, for all its acclaim, I suspect that few people today actually read *On Growth and Form,* and fewer still know quite what to make of it. Is it a work of literature or a work of science? Or perhaps, in the first instance, is it a work of religious faith, a testimony to the unity not only of nature but of human endeavor? The comments of Medawar and Hutchison suggest that they themselves were not quite sure. Exacerbating the ambiguity yet further is the book's heavy reliance on visual representations of structural analogies. The protozoologist Clifford Dobell described it "as a work of art no less than of science," and its frequent invocation as inspiration for exhibitions on the *Art of Form* suggests that Dobell was not alone in this view.[21] J. Arthur Thomson described it as "bewitchingly beautiful," but he also credited it as marking "a big advance in science," one that "will make other advances possible." He does not however say of what such advances might consist.

Clearly, *On Growth and Form* has been many things to many people, not least of all to Thompson himself. And because it is also a very large book, composed of many different parts, it has been possible for readers to draw quite selectively from its multiple offerings. For both of these reasons, it is impossible to give a simple account of its success. I will try instead to tease out some of the different strains of its appeal, while also probing the limits of that appeal. Doing so should help us understand how this book has persisted as a classic of scientific literature without obliging scientists to put it to use in their actual work, or, for that matter, even to read it; how it has remained, as it were, so robustly on the edges of science.

Two Possible Readings

Two possible readings come to mind immediately. The first is merely one more iteration of the age-old effort to explain the principles of living organisms by their similitude to mechanical structures—yet another variant of the design argument, in which the laws of biological function are inferred by analogy with the procedures by which mechanical devices are constructed. The second reading—which also has a long tradition—is that Thompson is demonstrating the unity of nature, much as Andreas Feininger sought to do in his evocative photographs on *The Anatomy of Nature* (1956).[22] The sense of the term "demonstration" here is strictly ostensive—achieved through the evocative power of visual similitude, without any pretense of offering scientific evidence. However, both of these readings seem to me to be off target and, as descriptions of Thompson's own intentions, seriously mistaken.

Thompson's method was certainly analogical, relying heavily on the evidence of similitude. But his principal point (despite his invocation of the engineer's crane) was to demonstrate—by such evidence—just how far one could go in the explanation of biological form without ever invoking *any* principle of design, or even, for that matter, of function. Indeed, as a morphologist, he saw structure as necessarily preceding function, not because function would automatically follow from form but because function (and this he firmly believed) was dependent upon form. His first need, therefore, was to explain the genesis of structure, and the point of many of his examples was to show that structure could be explained without any prior invocation of function or purpose.

Similarly, the arguments of *On Growth and Form* reached well beyond a mere exhibition of visual resemblances between the inanimate and the animate. Thompson had a methodology in mind, and he saw this procedure as conceptually modeled on the methods of physics. First, and most directly, his starting point was the identification of the physical conditions that would permit the construction of artifactual structures resembling those animate forms that survived and grew under similar earthly conditions. The actual constructions he described were generally not of his own making; they drew from the work of contemporaries (Stéphane Leduc, Carl Vogt, R. S. Lillie, and others) engaged in the artificial production of lifelike forms. The point of these efforts was not merely to imitate but to explain, to build an argument. For Thompson, as for Leduc and his colleagues, mimesis was the means rather than the goal. Their artificial constructions served as prosthetic links in a chain of argument. They were models of life exhibiting the adequacy of physical forces by their very existence, showing that such forces, under similar conditions as those obtaining for organisms, could—by themselves, without the guidance of architect or artisan—spontaneously give rise to the same (or similar) forms. Furthermore, insofar as these artifactual forms could, at least in principle, be explained by the mathematics of mechanics, then, Thompson argued, similar explanations ought also to be possible for biological forms.

Put in such general terms, it is not difficult to see what such a mode of argument might achieve. As the developmental biologist J. T. Bonner writes, eighty years after the book's first appearance, "It was D'Arcy Thompson who made us clearly see the importance of physical forces in the construction of living organisms. He showed that to a remarkable degree the form of animals and plants could be described in physical and mathematical terms; nature subscribed to the sound principles of engineering."[23]

Mechanics and Mathematics, Use and Beauty

Bonner, like Thompson himself, invokes the notion of "physical and mathematical terms" as if the relationship between the two were self-evident. Yet, when we examine the content of Thompson's demonstrations, a question very quickly arises concerning the actual role of mathematics in this work. Indeed, very little in the text is readily recognizable as mathematical; and should Thompson's own quite persistent references to the value of mathematical tools and concepts lead contemporary readers to expect either formal structures or equations, they are bound to be disappointed.

Dimensional analysis plays an important role in early discussions of limitations on the size, shape, and movement of organic bodies, and reference is made to the mathematical analyses of surface tension, minimal surfaces, capillarity, and stability by Raleigh, Maxwell, and others in later chapters (primarily for their application to the forms and stability of liquid drops, columns, and jets, of falling drops, vortex rings, coagulating solutions, and soap bubbles—in short, the kinds of objects that had fascinated a good number of mathematical physicists in the late nineteenth and early twentieth century). But the appearance of actual equations, or even of numbers, is surprisingly rare in Thompson's text. For the most part, the relevance of the theoretical analyses he invokes to the complex forms which most interest him (that is, the forms of actual biological organisms) is mediated by laboratory manipulations of liquid drops, columns, solutions—all of which were, like Leduc's osmotic growths, contrived to mimic organic forms. The resulting phenomena are presented not in mathematical but in visual form—sometimes in diagrams and graphs but more commonly in drawings and photographs. Why, then, the emphasis on mathematics, either in Thompson's own description of his project, or, for that matter, in posterity's description?

The final clause of Bonner's comment suggests an importance of Thompson's work not so much to the history of what we would today call mathematical biology but to the discipline we call biomechanics. Both of these disciplines have grown a great deal in recent years, but the overlap between them is remarkably slight.[24] Today's readers may be quick to distinguish between physics and mathematics, and even more so between engineering principles and mathematical concepts, but, interestingly enough, Thompson did not make such a distinction.

Much of Thompson's writing suggests that he rejected (and would perhaps even have been mystified by) such a divergence between biomechanics and mathematical models. Where we might read his mission as a dual one, he saw it as unified. In the early remarks that I quote here, in his easy and repeated alternation—"physical or mathematical," "explaining by geometry or by mechanics"—we find evidence of an underlying faith that bound the duality of his project into a unity. For Thompson, the interchangeability of mathematical concepts and dynamical principles, of geometry and mechanics, is justified in part by the assumption that such principles are necessarily grounded in mathematics and in part by the natural conjunction and even isomorphism he took for granted between use and beauty. This faith is made explicit in the Epilogue, where he writes: "I am no skilled mathematician . . . but something of the use and beauty of mathematics I think I am able to understand. I know that in the study of material things, number, order and position are the threefold clue to exact knowledge; that these three, in the mathematician's hands, furnish the first outlines for a sketch of the Universe . . . Moreover, the perfection of mathematical beauty is such . . . that whatsoever is most beautiful and regular is also found to be most useful and excellent" (p. 1097).

Yet elsewhere, we find him drawing on other traditions, traditions in which a distinction between "the use and beauty of mathematics" seems to be quite decisively marked. Thus, in the epigraph from Dr. Johnson's fourteenth *Rambler* with which Thompson introduces his first chapter (second edition), we find the invocation of what to us is the more familiar distinction between pure and applied science: "The mathematicians are well acquainted with the difference between pure science, which has to do only with ideas, and the application of its laws to the use of life, in which they are constrained to submit to the imperfections of matter and the influence of accident." But from Thompson's perspective, the appearance of contradiction here is largely illusory—an artifact of the multiple meanings of the word "use." Where Johnson refers to the application of mathematics to "the use of life," Thompson might be understood as referring to the application of mathematics to "the use of the mind." Mathematics shows "her peculiar power" in enabling us "to combine and to generalize," that is, in facilitating understanding of the highest sort, unhampered by "the imperfections of matter" (p. 1028). He may have faulted the founding father of morphology for ruling "mathematics out of place in natural history," but he was more than a little sympathetic to Goethe's criticism of the constraint exerted on man's ability to understand the world of nature by the "compulsion to bring what he finds there under his control" (p. 2). In his view, the best and highest uses of mathematics lay well beyond the range of any such compulsion; indeed, it was mathematics that would lead us along the path that Goethe himself had advocated—to a proper appreciation of the "variety of relationships livingly interwoven."[25]

Thompson's particular blend of diverse metaphysical traditions may have been idiosyncratic, but the component strands upon which he drew were well-established cultural resources, readily available for adaptation to his own agenda. On the one hand, he could draw on classical conjunctions of mathematics with truth and beauty and, on the other hand, on the equally venerable tradition that (at least since Descartes) bound mathematical concepts to physical principles, geometry to mechanics, number to utility. But almost certainly, the tradition in which the joining of physics with mathematics would have been of most immediate relevance (and surely most immediately compelling in the minds of the zoologists he saw as his audience) was the particular biological tradition against which he had set himself in opposition. The reluctance of his zoological colleagues to bring mathematics into biology—inherited from Bichat, Pascal, Goethe—applied equally to physics, and usually in the same breath. As Thompson explains it, it was a reluctance "to compare the living with the dead," and in this opposition between the living and the dead, mathematics, mechanics, and physics were seen as sitting together, all on the side of death.

But through the invocation of another tradition, Pythagorean if you will, he found a way to annul this insidious association, to make the conjunction of these two apparently irreconcilable realms more palatable, both to himself and to his readers. For in the identity between use and beauty, an identity that was itself wrought by "the perfection of mathematical beauty," lay something rather like redemption.

It goes without saying that the biological world that Thompson inhabited was very different from our own, shaped by oppositions now foreign to us. The distinctions today's scientists find necessary to draw lay elsewhere, and not surprisingly so: our struggles, and especially the struggles in which contemporary biology is engaged, are of an entirely different kind. Nevertheless, even now, we can still find Thompson's invocation of an ideal conjunction between the formal beauty of mathematics and the ambiguous promise of utilitarian value bewitching (its siren beckoning most conspicuously in contemporary theoretical physics), just as we can still find his evocative resonances between visual and formal beauty seductive. And this surely accounts for part of the persisting appeal of Thompson's work. But that such echoes can still be heard, and even responded to, should not obscure the magnitude of the epistemological divide between his time and our own. It is just this epistemological divide that needs to be examined if we are to understand why *On Growth and Form,* though a classic, has remained so resolutely on the margins of biological research.

"An Unusable Masterpiece"

Thompson's reviewers may have judged his work to be a masterpiece, but it is the silence of most practicing biologists—the virtual absence of any reference to this work in the experimental literature—which in the end has been more telling. This absence argues strongly for the conclusion that, in relation to the dominant research programs of twentieth-century experimental biology, *On Growth and Form* has proven, as Stephen Jay Gould once put it, "an unusable masterpiece."[26] While readers may have found great value in this work, it was evidently not the sort of value they could employ in their own research.

In his 1976 review of Thompson's work, Gould claimed that Thompson's "vindication has come only now," with the advent of the electronic computer. But the vindication Gould then saw was rather limited: he was referring to the revival by some recently published studies of morphological space of the argument Thompson developed in his final chapter, his theory of transformations, a method of coordinate grid transformations for relating different taxonomic forms. G. Evelyn Hutchison had earlier described this theory as "a floating mathematics for morphology, unanchored for the time being to physical science."[27] Gould argued, however, that new quantitative studies of morphological space—studies that (thanks to the computer) have only now become possible—provide the missing anchoring. Furthermore, over the last twenty-five years, a number of workers in theoretical morphology have continued to make use of Thompson's framework.[28] Yet, whether or not we would agree with Gould's claim, such efforts continue to remain on the margins of contemporary biological research. At best, they would have to be said to constitute a minor victory, and certainly not one that could count as a "vindication" of Thompson's overall ambitions.

Judging Thompson's success in terms of his own main agenda—that is, as an attempt to persuade his fellow zoologists of the utility of the methods of mathematical physics for the analysis of organismic development—we have to say that his success has been meager indeed. A scan of the technical biological literature shows little evidence of an impact on the main trends of biological research as these continued to unfold over the course of the century.[29] Thompson's efforts undoubtedly provided some satisfaction (and perhaps even ammunition) to contemporaries engaged in their own struggles against vitalism, teleology, and residual arguments from design; it is also clear that they provided satisfaction to later readers in their longing for an aesthetics of unity. But it seems equally evident—at least in hindsight—that such satisfaction did not follow even his most enthusiastic readers into their laboratory work, where more technically compelling concerns took clear priority, shaping not only their practical reasoning but also their styles of theoretical reasoning. And, as biological concerns came to be more and more driven by concrete experimental needs, it became ever harder to see how work of this sort could be of use.

Perhaps nowhere is the gap between Thompson's concerns and those of his successors more conspicuous than in his blatant disregard of genetics. An early critic of August Weismann's work, he saw Weismann's focus on the material units of heredity as yet another instance of the common over-preoccupation with the material elements of heredity and development. "The *things* which we see in the cell are less important than the *actions* we recognize in the cell."[30] To speak of "an hereditary *substance*" as responsible for development can only be justified "by the assumption that that particular portion of matter is the essential vehicle of a particular charge or distribution of energy, in which is involved the capability of producing motion, or of doing 'work'. For, as Newton said, to tell us that a thing 'is endowed with an occult specific quality, by which it acts and produces manifest effects, is to tell us nothing'" (p. 288).

In the first edition of his book, Thompson's reservations had been even stronger. There he wrote, "The question whether chromosomes, chondriosomes or chromidia be the true vehicles or transmitters of 'heredity' is not without its analogy to the older problem of whether the pineal gland or the pituitary body were the actual seat and domicile of the soul" (p. 286). But by 1942 the science of genetics had become far too well established to dispute wholesale, and Thompson was ready to be at least somewhat accommodating. Now he wrote, "The efforts to explain 'heredity' by help of 'genes,' which have grown up in the hands of Morgan and others since this book was first written, stand by themselves in a category which is all their own and constitutes a science which is justified in itself" (p. 340). Yet, "however true, however important" the results of those efforts may be, he opts even at this late date to "leave this great subject on one side"—not because he wishes either to dispute the hypotheses or to decry its importance "but because we are so much in the dark as to the mysterious field of force in which the chromosomes lie" (pp. 340–341). For his own purposes, that is, for providing a dynamic account of development, the concepts of genetics seem no to offer much help; accordingly, he chooses to ignore them altogether. As he explains at the end of Chapter 3:

> In this discussion of growth, we have left out of account a vast number of processes or phenomena in the physiological mechanism of the body, by which growth is effected and controlled . . . [We] have studied it as a phenomenon which stands at the beginning of a morphological, rather than at the end of a physiological enquiry. Under these restrictions, we have treated it as far as possible . . . on strictly physical lines. That is to say, we rule "heredity" or any such concept out of our present account, however true, however important, however indispensable in another setting of the story, such a concept may be.[31]

True to his promise, the only reference to genetic factors in the remainder of the text appears as a footnote to a discussion of "shrinkage patterns" in the formation of shells, vertebrae, and dried peas. Here he writes, "The difference between a smooth and a wrinkled pea, familiar to Mendelians, merely depends, somehow, on amount and rate of shrinkage" (p. 564).

We need only to imagine the response of a classical geneticist—or, for that matter, the bemusement with which most of today's readers would respond to such an explanation of phenotypic difference. The difficulty is not that Thompson was wrong in suggesting that the difference between a smooth and a wrinkled pea depends on the amount and rate of shrinkage (he may even have been right). The problem is rather, first, with the absence of even a proposal for an empirical test of this hypothesis and, second—and from the geneticist's perspective, especially egregious—with his willingness to assign sole or primary effective cause (his "merely depends") to such a mechanical process, solely on the basis of inference from the end result, and to relegate genetic factors to the no-man's-land of "somehow."

Geneticists clearly had another "merely" in view (for them, genes were the primary if not sole effective cause) and, as clearly, another use of the unexamined domain of "somehow." It was a part of their strategy—one might even say, of their disciplinary commitment—to consign the physical processes on which Thompson focused to their own no-man's-land of gene action.[32] For the purposes of their accounting, it seemed more appropriate to bracket (rule out) the physical and chemical dynamics by which a difference between wrinkled and smooth peas "merely depends, somehow" on the presence or absence of a particular genetic factor. One might say that the geneticists' focus was on the initial conditions observed to be necessary—the presence of particular genes (or alleles)—and that they were willing to leave questions about the sufficiency of such factors for realizing the final state of the organism (its phenotype) in abeyance.[33] Furthermore, unlike Thompson, geneticists had a flourishing experimental program to match their particular causal focus.

Both in its classical and its molecular phases, genetics may well be claimed as the most successful research program of twentieth-century biology. Yet it was far from the only such program; moreover, it has not always (and especially not in the first half of the century) occupied the dominant status that it does today. Taken as a whole, what most distinctively marks biological science in the twentieth century is its focus on experimental analysis, rather than on genetics per se. Accordingly, the absence of reference to Thompson's work in the technical literature needs to be understood in more general terms, and not just as a difference in causal focus. The larger problem, I suggest, reflects important epistemological differences between the cultures of mathematical physics on the one hand and experimental biology on the other. Furthermore, as the century wore on, these differences grew ever more marked and manifested themselves in radically different criteria for what is to count as "explanation."

Differences of Culture

By his own admission, the models Thompson presented were highly oversimplified, leaving "out of account a vast number of processes or phenomena in the physiological mechanism of the body, by which growth is effected and controlled." But this, after all, is just how the physicist proceeds:

> It is the principle involved, and not its ultimate and very complex results, that we can alone attempt to grapple with. The stock-in-trade of mathematical physics . . . is for the most part made up of simple, or simplified, cases of phenomena which in their actual and concrete manifestations are usually too complex for mathematical analysis . . . When we attempt to apply the same methods of mathematical physics to our biological and histological phenomena, we

need not wonder if we be limited to illustrations of a simple kind, which cover but a small part of the phenomena.[34]

The strengths that Thompson saw in the methods of mathematics and physics lay in their economy, in their focus on idealized examples, and in their pursuit of the "theoretically imaginable."[35] But experimental biologists have tended to see in these very characteristics not strength but weakness. Above all, they failed to see how the kinds of arguments Thompson offered might be usable for—or, for that matter, even relevant to—their own immediate concerns. From their perspective, it was difficult to see how Thompson's arguments contributed to what they would have regarded as a satisfying explanation. Often, his analyses did not even address the phenomena which they considered to be most in need of explaining. One might accept, as Bonner has claimed, that Thompson had shown how "the form of animals and plants *could* be described in physical and mathematical terms" (my emphasis) but still have to acknowledge that he had little to say about the particular processes by which the forms of individual animals and plants were in fact actually constructed. It is just these particularities of biological processes that have been the primary concern of experimental biology throughout the past century.

For Thompson, the goal of explanation appears to have been primarily one of sufficiency—in only a few cases did he argue for logical necessity, and virtually never for empirical necessity. He said in effect, this is how it *could* happen, not how it *need* happen,[36] and certainly not how it *does* happen in any particular instance. One might say that what he found most compelling about mathematics was not so much its deductive power as its power to lead our imagination away from the particular instances found in the real world and toward that which mattered most to him: the underlying essence of which the particular is a mere instance.[37] In a rather startling echo of Tennyson's depiction of nature's own mission, he wrote, "We must learn from the mathematician to eliminate and to discard; to keep the type in mind and leave the single case, with all its accidents, alone; and to find in this sacrifice of what matters little and conservation of what matters much one of the peculiar excellences of the method of mathematics" (p. 1032).[38]

The tension between such a mandate and more familiar conventions of biological practice was already evident when Thompson was preparing his first edition. In fact, his awareness of this tension led him to anticipate a form of opposition from many of his zoological colleagues (even those in organic morphology) that went far beyond any lingering repugnance "to treat the living body as a mechanism."[39] Here was an opposition based on epistemological and aesthetic leanings rather than on ontological commitments, and he sought to counter one of its primary sources head-on: "A large part of the neglect and suspicion of mathematical methods in organic morphology," he wrote,

> is due . . . to an ingrained and deep-seated belief that even when we seem to discern a regular mathematical figure in an organism . . . [it is] the details in which the figure differs from its mathematical prototype [that] are more important and more interesting than the features in which it agrees; and even that the peculiar aesthetic pleasure with which we regard a living thing is somehow bound up with the departure from mathematical regularity which it manifests as a peculiar attribute of life. This view seems to me to involve a misapprehension . . . We may be dismayed too easily by contingencies which are nothing short of irrelevant compared to the main issue; there is a *principle of negligibility.* Someone has said that if Tycho Brahé's instruments had been ten times as exact there would have been no Kepler, no Newton, and no astronomy.[40]

Thompson's advocacy of "a principle of negligibility" and the contrast he posed between mathematical regularity and biological irregularities, between lawfulness and contingency, between the "irrelevance" of individual differences and the greater relevance ("the main issue") of common denominators were explicitly addressed to the nineteenth-century "organic morphologist." His work might therefore seem to belong to another era altogether. The very designators of his areas of specialization (anatomy and morphology) and even of his professional position (chair of Natural History) recall categories that were effaced in the newer disciplinary taxonomies of the twentieth century. Biological science had progressed, and the routes along which it was to continue its advances neither overlapped nor intersected with the path he had envisioned. The major breakthroughs came from descriptions of component structures rather than from dynamical analyses, and they were achieved by focusing on just the sort of experimental particularities that Thompson's principle of negligibility would have had us ignore. At least until quite recently, the successes of experimental biology, and especially of experimental genetics, have only reinforced experimental biologists' suspicion of mathematical methods. With the notable exception of population biology (in which I include ecology, population genetics, and epidemiology), the gulf between "biological" and "mathematical" grew even wider as the century wore on. Thus, with only the slightest of modifications, Thompson's remarks might as well (if not even more aptly) have been addressed to the generations of experimental biologists who came in the wake of his ambitious undertaking, and perhaps especially to those concerned with the processes of development.

Conclusion

To the extent that Thompson's aim was to instill an appreciation for, and an interest in, the value of mechanical and mathematical models among his biological colleagues, it would have to be said that he failed. The factors contributing to this failure are many.

One is disciplinary competition between the biological and the physical sciences. Another is the fear and mistrust among experimental biologists engendered by their traditional lack of mathematical skills. A third is the relative absence of direct fit between the terms and concepts employed in such models and available experimental handles, and hence the difficulty in putting them to use in experimental programs. But as I will argue more fully in the next chapter, the paucity of interest among experimental biologists in mathematical models also reflects differences of a more basic nature—differences not only in skills, disciplinary status, or available tools, but in the values that determine what is to count as explanation, and even what is to count as knowledge.

Earlier, I quoted Thompson's own comments on differences in the relative importance attributed to contingency versus regularity, to differentiating details versus generalizable essences. And in comparing Thompson's explanatory goals with those of classical geneticists, I remarked on the different values placed on necessity and sufficiency. But necessity too is an ambiguous term, understood in mathematical culture as logical but in most of biological culture as primarily empirical. Just what should be the relative weight given to demonstrations of logical and empirical necessity? Logical necessity, after all, is only relative to one's starting assumptions, and how can we know that these are right? Finally, and closely related to this last point, are what must be called moral differences, differences that bear not just on what was to count as explanation but on what was (and is now) to count as properly scientific work—that is, on the kind of labor that can qualify as a warrant for true and certain knowledge. Can, for example, mere arm-chair theorizing—work that requires only paper and pencil and not the manual labor of actual experiment—serve as an adequate basis for epistemological entitlement? The judgment of most experimental biologists of the past century (and especially of those working in the United States) has been a decisive no.

Such differences, I claim, demarcate distinctive epistemological cultures in the practice of science. They underlay not only the fate of Thompson's own particular efforts to overcome "the neglect and suspicion of mathematical methods" but also the larger history of mathematical biology in the twentieth century.

CHAPTER THREE

Untimely Births of a Mathematical Biology

One considerable advantage both mathematics and mechanics may afford the naturalist, is, by schemes, figures, representations, and models; which greatly assist the imagination to conceive many things, and by that means enable the understanding to judge of them, and deduce new consequences therefrom.

Robert Boyle, *Some Considerations Touching the Usefulnesse of Experimental Naturall Philosophy* (1663–1671)

Ich behaupte nur dass in jeder besonderen Naturlehre nur so viel eigentliche Wissenschaft angetroffen könne als darin Mathematik anzutreffen ist.*

Immanuel Kant, *Gesammelte Schriften* (1786)

'Arcy Thompson is sometimes claimed as the father of mathematical biology, but, as is so often the case with the granting of retrospective paternity, such claims may have more to do with legitimization than with actual kinship. To be sure, Thompson himself invited such affiliation by his frequent emphasis on the importance of mathematics to biology; yet, conspicuous differences prevail between what the term mathematics meant to him and what it meant to his successors, and they require noting. By his own account, Thompson was no mathematician, and, as we have seen, his treatise contains few equations or even numbers. In fact, they may not even have been missed, for neither computation nor calculation appears to have been on his mind. His vision of mathematics

* "I maintain only that in every special doctrine of nature only so much science proper can be found as there is mathematics in it" (Kant, 1786).

did not depend on the invention of any sort of calculus; rather, it was an abstract formalism that would enable us to "discover homologies or identities which were not obvious before, and which our description obscured rather than revealed."[1] To him, the essence of mathematics was to be found in geometry rather than in either algebra or calculus, and its insights were to be represented in figures rather than in equations or numbers. As he wrote, "We are brought by means of it in touch with Galileo's aphorism (as old as Plato, as old as Pythagoras, as old perhaps as the wisdom of the Egyptians), that 'the Book of Nature is written in characters of Geometry'" (p. 1026).

By contrast, for those who have since attempted to develop a discipline of mathematical biology, calculus and the differential equation provided the principal instruments for the application of mathematics to biology, just as they had done for physics. Also, and not incidentally, these people were professional mathematicians and physicists rather than biologists, well schooled in both the methods and mores of mathematical physics. C. P. Snow had bemoaned the fracturing of intellectual life into two cultures, one scientific and the other humanistic; but when Snow was writing, the scientific world was itself a world divided, and the demarcations between and among its own subcultures were clearly recognizable. Perhaps primary among these for scientists of Snow's generation was the cultural divide between the mathematical and the biological sciences; and the fact that the major figures in the history of mathematical biology came from what was, for biologists, an effectively alien culture is of unmistakable importance to this history. Unlike Thompson, they were interlopers.

Indeed, whether or not D'Arcy Thompson merits the title of father of mathematical biology, most biologists would regard such an attribution as a dubious honor indeed. After decades of struggle, a professional society was finally founded in 1973.[2] Yet, despite repeated efforts over the course of the century to legitimate the formulation of mathematical models as a proper part of biology—with a function comparable to that which had been so clearly demonstrated in physics—mathematical biology has repeatedly failed to find acceptance as a rightful branch of the biological sciences. To be sure, the existence of its own journals, programs, and funding opportunities all indicate that this subject has succeeded in establishing itself as a discipline; however, its journals are read only rarely by practicing biologists, its programs are not housed in biology departments, and only a very small portion of its funding is provided under the rubric of biological science.[3] There are indications that this may now be changing (see Chapter 8), but until quite recently the history of such efforts—with the notable exceptions of ecology and population biology—has largely been one of frustration on the part of mathematical biologists and lack of interest (if not outright hostility) on the part of the vast majority of biologists.

To put it simply, one could say that biologists do not accept the Kantian view of mathematics (or rather, of mathematization) as the measure of a true science; indeed, they have actively and often vociferously repudiated any such criterion. Nor have practicing biologists shown much enthusiasm for the use of mathematics as a heuristic guide in their studies of biological problems. Perhaps nowhere has this impatience been more marked than among students of development and growth, particularly in the United States. To highlight the difficulties encountered by those who have sought to model these processes mathematically, and to try better to understand the difficulties they encountered, I begin with a brief discussion of the efforts—and ultimate failure—of Nicolas Rashevsky, a man widely castigated for his very presumption. I then turn to the contributions of Alan Turing in 1952, which, after an almost twenty year long hiatus, proved to be of signal importance to the reemergence of mathematical biology in the 1970s and to its establishment as a legitimate (even if not biological) discipline.

Nicolas Rashevsky

Nicolas Rashevsky (1899–1972) liked to describe himself as "a stubborn Ukrainian."[4] Born and trained as a theoretical physicist in czarist Russia, he immigrated to the United States in 1924, where he found employment as a physicist at the Westinghouse Research Labs in Pittsburgh. Here, while working on the onset of instability in liquid droplets, he began to wonder whether a similar mechanism might account for the division of biological cells. Taking this question to a biologist at the University of Pittsburgh, he was astonished to learn that almost nothing was known about the physics of cell division, and he set out to rectify this deficiency. Further inspired by reading Thompson's work, he conceived the ambition of building "a systematic mathematical biology, similar in its structure and aims to mathematical physics." Crediting Thompson with the idea in the preface to his own magnum opus, *Mathematical Biophysics: Physico-Mathematical Foundations of Biology* (1938), he wrote, "The timeliness of such an enterprise has been emphasized on several occasions, especially in the remarkable book, *Growth and Form,* by D'Arcy W. Thompson."[5]

A Rockefeller Fellowship brought him to the University of Chicago in 1934, and his institutional opportunity came the following year with an appointment in the Psychology Department at Chicago as an assistant professor in mathematical biophysics and, soon after, in the Department of Physiology. In 1938 Rashevsky published the first edition of *Mathematical Biophysics,* a compendium of his accumulated work of the preceding eleven years. "To bring mathematical biology to the same level [as mathematical physics]," he wrote, "is certainly not the work of any single individual. But somebody must start that task, must lay the first stone" (p. ix). One year later, he founded the *Bulletin of Mathematical Biophysics* as a vehicle for his own work, for that of his students and colleagues (such as John Reiner, Herbert Landahl, and Alston Householder), and for others with similar interests.[6] By 1940 he had succeeded in establishing an independent degree-granting program in mathematical biophysics overseen by a special "committee," and over the next quarter of a century he devoted much of his energy to establishing a new discipline by that name (or, after 1947, by the name mathematical biology).[7] His efforts peaked around 1953, but for more than another decade to come, the name of Rashevsky would remain, both in the United States and abroad, virtually synonymous with the discipline he had sought to found.

In 1954 everything collapsed: the budget of Rashevsky's Committee of Mathematical Biology (which Rosen estimates at thirty members in 1953) was drastically cut, Rashevsky's two grants from the National Institutes of Health were not renewed, and he was obliged to reduce his program to its bare bones.[8] With the arrival of a new university administration in 1960, funding was restored, but the reprieve proved to be short-lived. Another contretemps with the administration (this time, over the matter of Rashevsky's successor) led to his resignation a few years later, just a few months before he was due to retire, and by then both his program and his discipline were in disarray.

Today, little remains of either his institutional or scientific efforts, and his early contributions to the mathematical study of neural nets, morphogenesis, and cell division have been all but forgotten in the more familiar discussions of these subjects. In a brief history of the Society of Mathematical Biology, Michael Conrad writes that, with the successes of molecular biology, "a generation of theoretical 'speculation' was being sent to the graveyard in body bags."[9] Rashevsky was the first and most conspicuous casualty, and before long his very name had become a source of embarrassment—perhaps especially so to new generations who followed in his shadow.

The rise and fall of the Rashevsky school is testimony to many things: the strengths and weaknesses of his particular vision and personality, the vicissitudes of local and global politics, the tools he had available, and the state of biological science during this period. My focus here is limited to just one aspect of that history, namely, the conspicuous tensions between Rashevsky's ambitions and the prevailing ethos among experimental biologists of his time. Ever since his decline, the primary criticisms that have been leveled against Rashevsky are, first, his apparent disregard for the molecular and biologically specific material aspects of the problems he considered and, second, his failure to engage in any sort of interaction with practicing biologists. But dialogue is a two-way process, and some degree of responsibility for that failure lies with the biologists themselves.

In fact, he did make at least one effort to interest biologists in the kind of work he was doing, and quite early on. In 1934 he presented a preliminary version of his "physico-mathematical" analysis of the forces acting on an idealized (spherical) cell to an audience of biologists at the second meeting of the Cold Spring Harbor (CSH) Symposia on Quantitative Biology. This presentation provoked an encounter with Charles Davenport, E. B. Wilson, and Eric Ponder that is both informative and revealing, for here, in the responses Rashevsky evoked from his biological colleagues, one finds an unusually clear expression of some of the forces that were already at play, long before the advent of molecular biology, in the ultimate failure not only of his own program but of much of Thompson's more eclectic agenda as well. Indeed, I suggest that this encounter may even have been definitive.

The subject of the symposium was *Aspects of Growth,* and Rashevsky began by asking: "Do we need to assume some special independent mechanisms, which produce at a certain stage of the cellular life a division, or are those mechanisms merely the consequences of more general phenomena, which we know occur in all cells?"[10] His answer to this question was unambiguous: cell division, he argued, can be explained as a direct consequence of the forces arising from nothing more than cell metabolism. But Davenport, among others, was not happy. He retorted, "I think the biologist might find that whereas the explanation of the division of the spherical cell is very satisfactory, yet it doesn't help as a general solution because a spherical cell isn't the commonest form of cell. The biologist knows all the possible conditions of the cell form before division . . . In the special cases of egg cells and cleavage spheres, this analysis may prove very valuable. But after all, these are only special cases."[11]

Rashevsky's counter-response was far from reassuring. While it began reasonably enough, merely restating the traditional faith of the mathematician, albeit in a way that could only have fueled Davenport's discontent, it quickly escalated from a simple assertion of cultural difference to one of anticipated supremacy:

> It would mean a misunderstanding of the spirit and methods of mathematical sciences should we attempt to investigate more complex cases without a preliminary study of the simple ones. The generalization of the theory, to include non-spherical cells, is indeed needed, and this will be the subject of research after the simpler cases are thoroughly and exhaustively studied . . . To my mind it is already quite a progress, that a general physico-mathematical approach to the fundamental phenomena of cellular growth and division . . . has been shown to be possible. Judging by the development of other mathematical sciences, I would say that it will take at least twenty-five years of work by scores of mathematicians to bring mathematical biology to a stage of development comparable to that of mathematical physics. (p. 198)

But it was E. B. Wilson who had the last word. In a brief paper entitled "Mathematics of Growth" that directly followed Rashevsky's presentation, Wilson laid out the views of his colleagues regarding the place of mathematics in biology in a series of "axioms or platitudes":

I. Science need not be mathematical.

II. Simply because a subject is mathematical it need not therefore be scientific.

III. Empirical curve fitting may be without other than classificatory significance.

IV. Growth of an individual should not be confused with the growth of an aggregate (or average) of individuals.

V. Different aspects of the individual, or of the average, may have different types of growth curves. (p. 199)

Mathematics, Wilson concluded, may be helpful to the study of populations but not to individuals. And even at its most useful, it is "decidedly auxiliary to the matter of (1) finding out the facts about growth and (2) thinking out the variables which are important for growth" (p. 200).

In the discussion that followed, Davenport expressed his "cordial agreement" with Wilson, and Eric Ponder (Director of CSH and organizer of the symposia from 1936 to 1940) summed up the mood of the audience in a statement yet more pointed than Wilson's own:

> One point upon which there seems to be pretty general agreement is that there is little relation between the amount of which has been done on the mathematics of growth and the clarification of the subject which has resulted. As [James] Gray said some six years ago: "It is intrinsically improbable that the behaviour of a growing system should conform to that of a simple chemical system, and the conception of growth as a simple physico-chemical process should not be accepted in the absence of rigid and direct proof." Both Dr. Wilson and Dr. Davenport seem to be of the same opinion, and I

> question if the conclusion ought not to be put even more strongly. Work on the mathematics of growth as opposed to the statistical description and comparison of growth, seems to me to have developed along two equally unprofitable lines . . . It is futile to conjure up in the imagination a system of differential equations for the purpose of accounting for facts which are not only very complex, but largely unknown . . . It is said that if one asks the right question of Nature, she will always give you an answer, but if your question is not sufficiently specific, you can scarcely expect her to waste her time on you . . . What we require at the present time is more measurement and less theory . . . There is an unfortunate confusion at the present time between quantitative biology and bio-mathematics . . . Until quantitative measurement has provided us with more facts of biology, I prefer the former science to the latter. (p. 201)

Rashevsky returned to Cold Spring Harbor only once, to attend a symposium on neuronal activation in 1936. On that occasion, his respondents were mainly biophysicists (such as K. S. Cole, O. H. Schmitt, H. Grundfest, and A. M. Monnier), and he fared somewhat better. But the coolness of his reception by the geneticists and cell biologists at the earlier CSH meeting had left its mark, and it would have an effect that went far beyond him as an individual. Though Rashevsky had elicited this harsh response, Wilson's axiomatic distinction between the growth of an individual and the growth of an aggregate (or population), with its implicit (and Ponder's explicit) judgment that mathematical analyses of the former were "unprofitable," clearly applied more generally. Indeed, it was an attack aimed at the very core not only of his own vision of a mathematical biology but of D'Arcy Thompson's as well.

In 1940 Miloslav Demerec took over as director of the laboratories, explicitly steering the future course for research at CSH, and for its annual symposia, in the direction of genetics. Henceforth, the place of mathematics in these symposia would be delimited, as Wilson had recommended, to the analysis of the growth of populations rather than of individual organisms, that is, to its uses in population biology.[12]

Rashevsky clearly got the message. Recalling his experience at CSH, he tried once again to explain his strategy in his preface to the first edition of *Mathematical Biophysics:*

> Objection has been raised against the use of the word "cell" to describe a highly oversimplified conceptual system, which, true enough, possesses some properties of a living cell but lacks a much larger number of other properties. To this our answer is as follows: In the early days of the kinetic theory of gases the whole theory was built on the concept of a molecule as an elastic billiard ball. The present development shows us a molecule as a system of tremendous complexity, having scarcely any analogy to an elastic rigid ball. Of course, even in the older days, the creators of the kinetic theory were perfectly well aware that the concept of a molecule as an elastic ball was far from the actual truth. Yet they used it up to a point with great success.[13]

But he did not try again to interact directly with biologists who had so little appreciation for such a conception of theory.[14]

It has been suggested that his particular work on the physics of cell division stirred real interest in the 1940s at Oak Ridge, but those for whom it may have seemed useful were physicists and applied mathematicians working on nuclear fission, not biologists trying to make sense of fission in living cells.[15] His work on neural interactions fared better, but again, those who found it most compelling were early workers in artificial intelligence (for example, Walter Pitts, Herbert Simon, Marvin Minsky, and A. S. Householder) rather than neurobiologists per se. And of his work on the role of diffusion and metabolism in the generation of organic form, regarded by some as a precursor to Alan Turing's reaction-diffusion model (discussed below), little trace remains.[16] Turing's work was surely subject to many of the same criticisms as

Rashevsky's, and it aroused equally little interest among his biological colleagues. But unlike Rashevsky's contributions, it remains of interest for contemporary mathematical biologists to the present; indeed, this paper is often seen as having triggered the reemergence of the discipline.

Alan Turing on Morphogenesis

Alan Turing's name is not normally associated with developmental biology, but his 1952 paper, "The Chemical Basis of Morphogenesis," has had a deep and lasting influence on virtually all subsequent attempts to model these processes mathematically. Turing's foray into this quintessentially biological problem might come as a surprise, but from his own point of view the connection was clear: his interest in the problem of morphogenesis arose directly out of his preoccupation with the design of thinking machines, and particularly out of his curiosity both about how human brains were designed and about how the structures of that design came into being. As Turing's biographer, Andrew Hodges, writes, "Somehow the brain did it, and somehow brains came into being every day without all the fuss and bother of the minnow-brained ACE. There were two possibilities: Either a brain learnt to think by dint of interaction with the world, or else it had something written in it at birth—which must be programmed, in a looser sense, by the genes. Brains were too complicated to consider at first. But how did anything know how to grow?"[17]

An obvious place to begin was at the beginning, at the start of an organism's life—that is, with the fertilized egg. How "growth was determined was something 'nobody has yet made the smallest beginning at finding out'" (p. 430). By the early 1950s the science of genetics had certainly proven itself, but its very success also generated a certain frustration, at least among those who had remained troubled by the absence of an account for the reliable development of the characteristic form of a biological organism from one generation to another. On that question, genetics was not only silent but left a gaping conceptual hole, or, as some saw it, a fundamental paradox dividing genetics from embryology. The "one-gene, one-enzyme" hypothesis of Beadle and Tatum was surely an inspired contribution to part of the question of how genes acted, but it offered little if anything toward a solution of the problems of differentiation and morphogenesis: If all the cells of an organism have the same genes, and hence the same enzymes, how is one to account for the development and organization of the many different kinds of cells required for the characteristic structure and form of a complex organism? How could one bridge the gap between genetics and ontogenetics?

Turing, it seems, was determined to find an answer to this question; once and for all, as he told his friend and former student Robin Gandy, he meant "to defeat the Argument from Design" (p. 431).[18] The specific purpose of his paper, as he wrote in the abstract, was "to discuss a possible mechanism by which the genes of a zygote may determine the anatomical structure of the resulting organism."[19] In one sense, his timing may even have been opportune. By then, frustration over the limits of genetic analysis had reached something of a climax, and at least an echo of this frustration can be seen in the spate of publications appearing at that time on the problem of form in biology.[20] Of course, such a venture would inevitably take him away from the specific problem of brain structure, but, as he wrote to J. Z. Young, the one problem "is not altogether unconnected with the other problem. The brain structure has to be one which can be achieved by the genetical embryological mechanism, and I hope that this theory that I am now working on may make clearer what restrictions this really implies."[21]

True to form, Turing approached the task on his own, consulting scarcely any of his more biologically informed colleagues and citing only three biological works in his references (D'Arcy Thompson, C. M. Child, and C. H. Waddington).[22] No reference to Rashevsky's previous work appears, or to an even earlier and more directly relevant paper by Kolmagoroff and his colleagues in 1937, and the likelihood is that he was unaware of these precursors.[23] Nor for that matter did he rely on his own previous work on computer programs encoded in sequences of digits and inscribed on a linear tape. In fact, his approach to the problems of morphogenesis, centering as they did on continuous processes distributed in space, represented a radical departure from his thinking about computers. His basic idea was that "a system of chemical substances, called morphogens, reacting together and diffusing through a tissue, is adequate to account for the main phenomena of morphogenesis."[24]

The paper opens with a more general program for modeling embryogenesis (Turing calls it "a mathematical model of the growing embryo"). As he freely admits, the "model" is "a simplification and an idealization, and consequently a falsification" (p. 37).[25] This, however, does not bother him any more than it had Thompson or Rashevsky; like these earlier authors, he expresses the hope "that the features retained for discussion are those features of greatest importance" (p. 37).[26] The first challenge to be met in such a program comes from the complexities of cell structure and the obvious difficulties these pose for mathematical analysis, and Turing suggests two ways of dealing with this problem: either to treat cells as idealized points of a lattice (similar to the Ising model for ferro-magnetism, and a clear precursor to what have since become known as cellular automata) or to ignore them altogether and treat the embryo as a continuous structure. In both cases, the embryo is to be represented by a state function, just as would be conventional for theoretical descriptions of physical systems. He then divides the variables assumed to define the state of the system (in both its discrete and its continuous form) into two kinds: one, mechanical, and the other, chemical. For the discrete form of the model, mechanical variables include mass, position, velocity, and elastic properties of each cell, and the forces between cells; in the continuous form, they include stress, fluid velocity, density, and elasticity. The time dependence of these mechanical variables is given by Newton's laws of motion. The concentration of the constitutive chemical substances and their coefficients of diffusion make up the second set of variables, and these are assumed to be governed by the respective equations for reaction kinetics, obtained from the law of mass action and diffusion. The basic presumption of this description is that the interaction between chemical and mechanical variables will determine the growth of the embryo.

Even in this idealized description, however, the problem is still one of "formidable mathematical complexity"; moreover, while equations determining the two kinds of variables separately are known, apart from the possible effects of osmotic pressure, little is known about their interaction. Turing thus simplifies his model yet further, abandoning the growth of the embryo and restricting his considerations to the chemical dynamics of non-growing tissues. As he will show, these are sufficient for generating inhomogeneous patterns in the spatial distribution of chemical substances, and the implicit assumption is that the mechanical variables will follow suit, building on these chemical patterns to shape the physical structure of the growing embryo.

Indeed, this assumption is built into his analysis by the very term he coins: "These substances will be called morphogens, the word being intended to convey the idea of a form producer" (p. 38). What is a morphogen? Not something that has been either chemically or biologically defined, it "is simply the kind of substance concerned in this theory." In fact, anything that diffuses into a tissue and "somehow persuade[s] it to develop along different lines from those which would have been followed in its absence" will qualify. Turing suggests that "the evocators of Waddington provide a good example," or hormones, or perhaps skin pigments (p. 38). Genes too might be considered as morphogens, but because they do not diffuse, they form "rather a special class." Furthermore, since the role of genes is presumably purely catalytic, influencing only the rates of reactions, unless one is interesting in a comparison of organisms, they "may be eliminated from the discussion."

Thus simplified, Turing now has a model that is amenable to analysis. Postulating the presence of two "morphogens," X and Y, with diffusion constants of D_x and D_y, coupled by a system of hypothetical chemical reactions, it can readily be represented by a pair of coupled differential equations:

$$X_t = f(X, Y) + D_x \nabla^2 X$$

$$Y_t = g(X, Y) + D_y \nabla^2 Y$$

where f(X, Y) and g(X, Y) describe the effects of the chemical reactions between X and Y. Because only nonlinear reactions are capable of producing the kind of dynamics Turing is interested in, f(X, Y) and g(X, Y) must correspondingly be nonlinear functions.

More than fifty years before, H. Bénard had described the spontaneous formation of regular hexagonal convection cells ("tourbillon cellulaires") when thin layers of fluid are uniformly heated from below, and in 1916, J. W. S. Rayleigh, by analyzing a linearized form of the relevant fluid equations, explained this as a bouyancy-driven instability.[27] Rayleigh's analysis of the phenomenon (now referred to as Rayleigh-Bénard convection) quickly became the point of departure for applied mathematicians interested in explaining the formation of patterns in fluids of any kind, and so it was for Turing as well. The principal point, as Turing explained, is that "such a system, although it may originally be quite homogeneous, may later develop a pattern or structure due to an instability of the homogeneous equilibrium, which is triggered off by random disturbances."[28]

D'Arcy Thompson had also been interested in such phenomena, and he devoted a great deal of attention to them, seeking in the conditions of instability the explanation both of Leduc's artificial tissues and of the formation of real biological tissues. But the significant difference here is twofold: Turing now added chemical reactions to the forces of diffusion, and he both formulated and analyzed the equations for the process. How are the particular reactions between the two hypothetical morphogens to be chosen? To do so by looking at actual reactions "would settle the matter finally, but would be difficult and somewhat out of the spirit of the present inquiry" (p. 43). Instead, they are to be imaginary reactions, chosen first by the criterion that they will generate instability and subsequent pattern formation, and second by the criterion of simplicity. As he wrote, "It should be emphasized that the reactions here described are by no means those which are most likely to give rise to instability in nature. The choice of the reactions to be discussed was dictated entirely by the fact that it was desirable that the argument be easy to follow."

Since the resulting equations are nonlinear, the standard procedure begins with the identification of all homogeneous, equilibrium solutions (that is, where both $f(X,Y)$ and $g(X,Y) = 0$), and follows with analysis of the (readily solvable) linear equations governing small departures from equilibrium. But such a procedure tells us only where instability will occur and how rapidly departures from equilibrium will grow. To get at the particular forms of the non-equilibrium solutions, analysis of the full nonlinear equations is required. Ironically, computers capable of doing the job were not yet available to Turing, and he was obliged to rely on the tedious process of manual calculation. But even with nothing more than a handheld calculator, he was able to produce a number of striking results. He found that, on a ring, his equations would produce patterns of the sort needed for the "tentacle patterns of Hydra and for whorled leaves"; on a flat surface, they yield dappled patterns resembling those seen on a cow's hide or a leopard's skin; and on the surface of a three-dimensional sphere, the resulting patterns bear at least some resemblance to the gastrulation of an embryo—close enough, in any case, to convince Turing that "such a system appears to account for gastrulation" (p. 37).

The Imaginary and the Real: A Clash of Scientific Cultures

In this description of Turing's effort, we are presented with a veritable caricature of the mathematical physicist, one in which the contrast with the features of the experimental biologist are depicted in even sharper opposition than was possible to see from the brief portrait of Rashevsky. The contrast is helpful, for it serves to bring the clash of scientific cultures into clearer focus. But Turing's foray into biology lends additional clarity to this discussion of cultural difference for an even more obvious reason: his very stature effectively removes the complicating factors of personal and scientific mistrust that have so confounded the history of Rashevsky's efforts. The crucial question, here as for Rashevsky, is one of biological relevance, and this question has haunted Turing's analysis ever since the paper was first published. In this case, however, the lineaments of controversy emerge more distinctly just by virtue of being unmarred by any hint of aspersion on the author's scientific standing. In other words, Turing's status in the history of science obliges us to take his reasoning for biological relevance seriously. But just what was the character of this reasoning?

Two aspects are immediately striking, both of which bear on the nature of modeling and both of which warrant our attention. The first point to be noted is that Turing, like Rashevsky, seems to take the mere fact of a resemblance between the computed solutions and the phenomena being modeled—a resemblance that is at best conspicuously crude—as sufficient to render the model "adequate to account for" the phenomena in question. Second, despite his acknowledgment that his model for morphogenesis is "a simplification and an idealization, and consequently a falsification," on the assumption that it contains the "features of greatest importance," Turing concludes by suggesting "that the imaginary biological systems which have been treated, and the principles which have been discussed, should be of some help in interpreting real biological forms" (p. 72).

Questions about which features of a phenomenon are of greatest importance are bound to arise, for consensus on this crucial matter is never something that can be taken for granted. Research in biology, as in any scientific endeavor, depends on the exclusion (or bracketing) of many facets—sometimes of even conspicuously evident facets—of biological development, and researchers often disagree in their judgments of what to retain and what to exclude. But such differences, which are disagreements over the relative importance of the many different observable aspects of biological processes, are generally *internal* disagreements; they remain firmly embedded within the culture of the discipline. To be sure, Turing's willingness to eliminate genes from his discussion and to either ignore the presence of cells or reduce them to points on a lattice would have seemed odd to most practicing biologists of the time, and, to many, might even have suggested a positive affront. But in 1952 genetics had not yet achieved the dominance in biological thought that it was later to assume, and a significant number of biologists could still be found whose work focused neither on genes nor on cells. By far the more serious problem that Turing's model posed for biologists, like that posed by both Thompson's and Rashevsky's earlier efforts, was of another kind altogether, and it had to do with the explanatory value of models which rely on imaginary constructions and make no pretense to literal truth. And on this, consensus tended to override all internal disagreements.

Andrew Hodges, also a mathematician, refers to Turing's methodological remarks as "a classic statement of the scientific method."[29] And almost surely, most mathematical and physical scientists would agree. The notion that a model which is admittedly a fiction can, despite its fictionality, nonetheless capture the features "of greatest importance" and hence can serve a useful explanatory function has a long and esteemed tradition in these sciences. One might even say that models in the physical sciences are fictions by definition (or, as Hans Vaihinger would have called them, "semi-fictions"): they are analogical rather than literal, corresponding to an actually occurring phenomenon in some respects but not in others.[30] Furthermore, the sense in which they may be seen to correspond may be quite abstract, involving no literal match at any point. But what may well be of greatest importance here is the fact that they are made up, based on imagination rather than on observation.

The explanatory value of such fictional constructions in the physical sciences has been discussed at great length in the philosophical literature, and for many of these authors the reflections of James Clark Maxwell provide a *locus classicus*.[31] While still a young man, Maxwell argued that the electric potential should be seen as such a fiction: "We have no reason," he wrote, "to believe that anything answering to this function has a physical existence in the various parts of space, *but* it contributes not a little to the clearness of our conceptions to direct our attention to the potential function *as if* it were a real property of the space in which it exists."[32] Later, as he struggled to make sense of the properties of the electromagnetic field, he devoted much effort to the development of mechanical models of a medium (involving molecular vortices and connecting idle-wheels) that would be able to support electromagnetic waves. And here again, he argued that such models were to be taken in the sense of *as if* constructions. As he explained in a final summary of his views, "The attempt which I then made to imagine a working model of this mechanism must be taken for no more than it really is, a demonstration that mechanism may be imagined capable of producing a connexion mechanically equivalent to the actual connexion of the parts of the electromagnetic field."[33]

Instances of models as "semi-fictions" from the twentieth century are equally easy to find: we might think, for example, of the Bohr atom, or the nuclear shell model. Both of these models were clearly understood to be imaginary constructs, useful solely for their value in computing properties corresponding to those that can be experimentally observed (at least under appropriately crafted and regulated conditions). So common are such examples that Nancy Cartwright suggests, as a general principle, that models play a role in physics comparable to that of fables in the moral domain—they "transform the abstract [of physical laws] into the concrete [of experimental observations]."[34]

Turing's fictional construction for accounting for morphogenesis clearly falls within this tradition, but with a difference. Because the properties that can be computed from Turing's model are expected merely to bear *some* resemblance to the phenomena that are actually observed, and not required to literally correspond with any of their specific properties, it is both more abstract and less tethered—and hence, more remote from the real world—than these familiar examples from physics. In the spirit of

Maxwell's argument for the value of the potential function in studies of electricity, its value might be said to lie in "contributing . . . to the clearness of our conceptions"; it serves to direct our attention to the role of reaction and diffusion, *as if* the particular reactions he imagined were a real property of organisms as they exist, but in addition, *as if* the gap between calculated and observed properties can, at least provisionally, be ignored. Yet even with this difference, it also serves much the same function as the system of idle-wheel particles Maxwell hypothesized in his mechanical models of the ether, namely, as "a demonstration that [a] mechanism may be imagined capable of producing" the effect required.

Physics, however, is not biology. To the vast majority of biologists, such a conjunction between fiction and explanation would have (then, as now) appeared as a manifest contradiction in terms. What possible value could there be to an explanation posited on a purely imaginary system of chemical reactions, one for which not only is no evidence provided but, worse yet, for which the gathering of evidence is considered not even worth attempting? Turing was not suggesting that his hypothetical reactions corresponded to any real reactions occurring in the cell: it was mathematical fruitfulness and accessibility that guided their construction rather than any attempt at realism. His model had the virtue of being both simple and elegant, but it was at best biologically naïve and at worst irrelevant.

To be sure, it could be seen as a guide to a kind of interaction that *might* be occurring, but what exactly was the point of that? Even if organisms could build patterns (or form) in this way, the question for biologists is not could they but do they? Turing's model aimed at providing an in-principle (or how-possible) answer to a question that he saw as fundamental and that, despite its having receded from the immediate attention of most experimental biologists of the time, would still have had to be acknowledged as one of the main questions of biology. His reaction-diffusion mechanism was not only a possible answer to the question of how structure could arise in an undifferentiated system, but—given his two key starting assumptions of (a) genetic conservation under cell division, and (b) the absence of causally effective inhomogeneities in the egg's cytoplasm (both of which assumptions were fairly routine among geneticists at the time)—it was also the first physically plausible answer to this question to have been proposed.[35] Nevertheless, it evoked little interest in the biological community. As Eric Ponder had earlier summed up the attitude of his colleagues at Cold Spring Harbor, "It is futile to conjure up in the imagination a system of differential equations for the purpose of accounting for facts which are not only very complex, but largely unknown."[36]

How are we to understand this conviction of futility? One way would be to suppose that the absence of a plausible explanation, even of so basic a phenomenon as morphogenesis, was simply not troubling and therefore that the offer of a theoretically imaginable account, based merely on speculations about how the process *might* occur, met no felt need. Or alternatively, one might argue that such a need was certainly felt, but what was lacking was the confidence that it could be satisfied by mere theoretical imagining; after all, isn't one of the primary lessons of biology that of nature's capacity to outwit us?[37] If so, the question of hubris becomes crucial: just how much confidence is it appropriate to have in the powers of our own intelligence?

But either way, whether the difference in views reflects one of felt need, of hubris, or of both, it weighed sufficiently to mark a radical divide between the cultures of the mathematical sciences and those of experimental biology. The reception of Turing's model among the majority of practicing biologists of his time clearly suggests that, in the middle part of the century, the same divide that was evident in the time of Thompson and Rashevsky was still operative.

Causal Dynamics and Explanatory Satisfaction

Another feature of Turing's model portending problems in its reception needs to be noted as well. This too pertains to the issue of cultural difference, but it centers more directly on the question of explanatory satisfaction. Rather than reflecting differences in the value placed on imagination and observation, it bears precisely on the matter of causal expectations, and hence it reaches beyond the divide between mathematical and biological science. Causal expectations obviously depend on one's understanding of the meaning of the term "cause"; but because they are so closely tied to the question of what provides explanatory satisfaction, they also reflect something we might think of as epistemological aesthetics. Let me explain what I mean.

The usual assumption in the literature on explanation is that the primary task of a scientific explanation is to provide a causal account of a phenomenon. But what, we need to ask, is to be understood as a causal account? For many people, the notion of cause implies a motive force emanating either from some pre-existing material entity or entities (such as germs that cause disease, or genes that cause the appearance of traits) or from some precipitating event (in the sense that flipping a switch causes a light to go on). Accordingly, the expectation of a causal account is that it will identify the agent or event responsible for the effect. Does Turing's model provide such an account, even in principle? If so, to what does his analysis point as the locus of cause?

Here, the form of an organism is self-generating and self-organizing, arising *de novo* in each generation out of the dynamical interactions among the chemical products which the genes had (somehow) catalyzed into being. Genes are assigned causal responsibility for producing the relevant players but not for the subsequent generation of spatial structure. In fact, no locus of cause—no prior agency or pre-existing determinant—is invoked in this account of morphogenesis: causal responsibility is assigned not to particular material entities or events but rather to a set of interaction dynamics that might or might not be sensitive to the particular players involved. For those who expect an explanation to identify particular causal loci, such an account is *a priori* unsatisfying. Yet for others, it is the attribution of causal responsibility to particular entities or events that often appears unsatisfying, even to beg the question—especially when the role (or even presence) of such entities or events seems itself to require explanation. And for many problems in developmental biology—possibly even that of embryogenesis (at least as Turing posed that problem)—this is surely the case. From such a perspective, Turing's analysis is appealing precisely because "it offered a way out of the infinite regress into which thinking about the development of biological structure so often falls. That is, it did not presuppose the existence of a prior pattern, or difference, out of which the observed structure could form. Instead, it offered a mechanism for self-organization in which structure could emerge spontaneously from homogeneity."[38]

I wrote the words just quoted almost twenty years ago. They appear in a reflection on the communication difficulties that I had personally encountered in the late 1960s when trying to explain the virtues of a mathematical model that Lee Segel (an applied mathematician) and I had developed to explain to practicing biologists the onset of aggregation in cellular slime mold.[39] Our model had been deeply influenced by stability analyses in fluid dynamics, and, as we saw it, it served a strikingly similar function to that of Turing's model: it explained the emergence of structure out of the dynamical interactions (in this case, largely known) among cells that were initially undifferentiated, and thus it dispensed with the need to hypothesize the prior existence of some sort of pre-differentiated founder cells.[40] But the biologists with whom I talked saw no virtue whatsoever in our account; in fact, they clearly preferred the hypothesis of founder cells, despite the absence of supporting evidence and despite the absence of any explanation of how such specialized cells might themselves have arisen.[41] In an attempt to make sense of what, to me, was so utterly perplexing a response, I argued for a conceptual gap that went well beyond terminology, reflecting far more than a difference in understanding of the meaning of cause.

Explanatory satisfaction, I suggested, is akin to narrative satisfaction: the explanations that propitiate our need for understanding, the stories we like to hear, are those that meet the expectations we bring with us. Such expectations are formed from a reservoir of experiences that are not only technical and scientific but also social and political.[42] I also suggested a disciplinary component to the difference in perspectives: "Thinking like a mathematician and not like a biologist, I found it natural to look for a system that . . . might provide a demonstrable instance of such self-organizing principles."[43] Yet, while mathematicians (and now computer scientists) certainly have extensive experience with such narratives and might therefore find them more gratifying than do many non-mathematicians, familiarity is clearly not the whole story.[44] Indeed, based on what we know of his personal biography, I would hazard the conjecture that such an account may have had particular aesthetic appeal to Turing just because it depended so little on such pre-existing entities as genes and cells and hence owed so little to the influence of one's progenitors.

Reception of Turing's Model

Turing's attempt to generate form out of nothing more than a hypothetical set of coupled nonlinear chemical reactions and the diffusion of their products was slow to gain recognition from any quarter. A scan of the literature shows that, for the first fifteen years following its publication, Turing's work was cited on average just over two times per year. One of the few early responses—

and, to my knowledge, the only one from either embryology or experimental genetics—came from C. H. Waddington. But in many ways Waddington himself was an outlier, and it would be difficult to take him as representative of any particular scientific culture. His insistent straddling of both genetics and embryology is one sign of his anomalous position; his long-standing interest in theoretical biology, another.[45] In the 1930s, he had been part of a small but exclusive Theoretical Biology Club in Cambridge (together with Joseph Needham and J. H. Woodger), and in the late 1960s and early 1970s, he organized a series of conferences at the Villa Serbelloni under that name.[46] In fact, he was one of the few scientists of his generation able to negotiate between the explanatory values of mathematicians and those of experimental biologists, and it is just this (multiply) ambiguous cultural location that renders his response of particular interest here.[47]

Waddington well recognized the absence of an account in biology for the growth of biological form, and he regarded the lack as being first and foremost a theoretical one. It is thus hardly surprising that Turing's efforts immediately attracted his attention. In a letter to Turing sent less than a month after the paper appeared, he wrote, "It is very encouraging that some really competent mathematician has at last taken up this subject."[48]

Yet even Waddington's enthusiasm was measured. Although he found Turing's arguments "extremely interesting and suggestive," he immediately followed with the remark, "I rather doubt, however, whether the kind of processes with which you were concerned play a very important role in the fundamental morphogenesis which occurs in the early stage of development." He continued, "The most clear-cut case of your type of mechanism seems to me to be in the arising of spots, streaks and flecks of various kinds . . . such as the wings of butterflies, the shells of moluscs [sic], the skin of tigers, leopards, etc."[49] In short, Turing's mechanism may apply to pattern formation but not to embryogenesis.

The following year, Waddington went public with his reservations. On the one hand, he granted that "the demonstration that patterns may arise as a consequence of chance variations around an equilibrium is . . . an important one. It probably finds its best exemplification in such phenomena as the colour patterns on butterflies' wings, animals' coats and such things."[50] But to Waddington, who had thought long and hard about the dynamics of early development, one problem with Turing's model was immediately obvious: "It is only rarely that the student of higher organisms is faced with the problem of accounting for the arising of a pattern *de novo* in a previously homogeneous system. Much more commonly, the phenomenon which confronts us is the gradual elaboration of an originally simple system into a more complex one. Certainly the egg, from which individual development starts, is very seldom, if ever, a featureless sphere" (pp. 123–124).[51] Finally, in a remark that was to prove equally trenchant, he noted that "biological patterns are often expressions of dynamic equilibrium"; "not a static affair which is formed once for all and is merely there; on the contrary, it can be shown for many patterns that they are actively maintaining themselves all the time" (p. 124). Turing, he thought, may be "going too far" (p. 122). His model may prove useful for understanding pattern formation but probably not for understanding morphogenesis.

In any case, 1953 was hardly a propitious time for thinking for that problem, and even less for thinking about either the utility or the limitations of any such model. Though 1952 might have appeared as fortuitous timing for Turing's paper, the subsequent history of biology has proven otherwise. Less than one year later, and just six months after the appearance of Waddington's comment, came Watson and Crick's triumphal announcement of their resolution of the structure of DNA, with all of its well-known implications. Inevitably, DNA, translation mechanisms, and codes took center stage, and even the attention of mathematicians and physicists looking for greener pastures (or for new fields to conquer) was captured by the new excitement. The audience upon which Waddington's response might have had an impact had other things to think about. The net result was that, until the late 1960s, few readers from either culture (mathematical or biological) even noticed Turing's paper.

The intervening years were particularly fallow ones for mathematical biology, or, more accurately, for anything called by that name. In a recent book, Lily Kay reviews an extensive body of theoretical and mathematical work from the early days of molecular biology, but none of this work was referred to as "mathematical biology."[52] To many, at least in the United States, the very term had become contaminated by its association with Rashevsky, now somewhat notorious among biologists for what they saw as his cavalier disregard of experimental realities. DNA—with the discovery, first, of its structure and, later, of a correspondence between nucleotide sequence and amino acid sequence—delivered a powerful take-home message, for it suggested mechanisms for replication and protein synthesis, two of the central mysteries of biology, that not only had never been but probably would never have been theoretically imagined. Those theoretical (and/or mathematical) challenges that were stimulated by the unique properties of DNA were seen as being of a different order, if only by virtue of being grounded in experimental reality, however new that reality might be.

In ways that have been abundantly documented, the advent of DNA was transformative for a wide range of biological questions. Here however, in these debates, it served as a clear confirmation of traditional biological values. It demonstrated once again the conspicuous gap between logical possibility and biological actuality—a gap too large to be bridged by the investigation of ideal models. Organisms solve the problems they face with little regard for elegance, efficiency, or logical necessity; furthermore, the range of possibility from which they draw exceeds by far the range of human imagination, however talented and ingenious. And this lesson was taken to heart by even the most mathematically inclined of the new recruits.[53]

After 1953 Rashevsky found that recruiting new students to his program at the University of Chicago was more difficult than ever; even holding on to the students he already had was proving to be a challenge. By the 1960s, most of the early recruits from the physical sciences had been effectively reschooled—not only in new techniques but also in a new set of cultural values—and the use of "paper tools" in molecular biology was all but abandoned. Thus the particular window of opportunity that had been opened in the 1950s to enterprising young physicists and mathematicians was now closed. But for all its prowess, molecular biology had little to say about the problems that had preoccupied Thompson, Rashevsky, and Turing. The newly discovered mechanisms of replication and protein synthesis could not, by themselves, resolve the enigma of morphogenesis, and the central question of how biological form develops remained unanswered. Furthermore, American universities were in the meantime producing a steadily increasing number of Ph. D.'s in mathematics and physics, all of whom were in need of employment. Equipped with the tools of their home disciplines, a number of these set out in search of new fields to plow. Perhaps it was therefore only a matter of time before the all-but-abandoned vision of a mathematical biology would be resurrected; and Turing's efforts, ungrounded though they may have been, would likely resurface as at least a possible handle on this central problem of biology still waiting in the wings. Here was a problem waiting to be solved, and what could possibly be more attractive to trained problem solvers looking for work?

The Reemergence of a Mathematical Biology

The turnabout for Turing's contribution came in the late 1960s, first, with its promotion by Ilya Prigogine as an instance of dissipative structures and, at roughly the same time, with its rediscovery by a new generation of mathematical biologists.[54] Only then does one begin to find a significant measure of interest: the frequency of citation rises to almost eighteen per year in the early 1970s, and continues to climb steadily thereafter. In a recent review, J. D. Murray has described Turing's paper as "one of the most important papers . . . of this century."[55] An undeniable classic for mathematical biologists today, it remains to many an exemplar for their field. Over the last thirty years, it has given rise to a minor industry of reaction-diffusion studies in chemical and biological systems, most of which appear in specialized journals, and there are some indications that now it is even beginning to attract the interest of a few mainstream biologists.[56] Nevertheless, the question stands: Where, and for what, is its value seen to lie?

The three principal categories into which applications of Turing's model fall, listed in order of decreasing frequency, are: (1) spatial and temporal patterns in chemical systems; (2) the colored patterns of butterfly wings and animal coats; and (3) the development of form in early embryogenesis—Turing's initial focus of interest.[57] In each of these areas, the lessons of the preceding years were taken to heart. Unlike their predecessors, the new recruits made substantial efforts both to find experimental pegs on which to anchor their analyses and to interest experimentalists in testing their models. Furthermore, and especially in the United States where they were eager to distance themselves from the now discredited Rashevsky tradition, they avoided his label (mathematical biophysics) and opted instead for the terms biomathematics and mathematical or theoretical biology. By a number of indicators, they clearly prospered. New societies were created, new programs established, new journals launched.[58]

Yet despite such apparent marks of success, and notwithstanding a number of conspicuous efforts to ground their work in experimental realities, to make concrete predictions, and to enter into collaboration with experimentalists, few working biologists responded.[59] The initiative for new courses and new programs came primarily out of departments of mathematics or applied mathematics, and, except for their occasional incorporation into population biology and ecology, there they have remained. The specialized journals that had been either newly established or revamped may have flourished, but they attracted few readers from experimental biology. Similarly, those authors who sought to apply their theoretical analyses to available biological data often (in fact, usually) found the more traditional journals unreceptive. Chemistry, where their efforts were welcomed in the established journals and where the challenge of bridging theory and experiment was embraced by a number of experimental chemists, tells a notably different story.[60] Here was a problem seen to fall well within their own concerns, whether or not it might prove to be a prototype for biological processes.

By far the greatest disappointment has been with efforts to apply Turing's reaction-diffusion model to concrete (and clearly recognizable) problems in developmental biology. While scores of articles with models for phenomena of considerable interest to researchers in this field (for example, slime mold aggregation and development, chemotaxis, *Drosophila* embryogenesis) have been published over the last thirty years, only a handful of these appeared in journals that developmental biologists might actually read. Nor, when called to their attention, were they much interested. And why would they have been? Molecular biology may not have provided a handle on the particular problem of morphogenesis, but it left biologists with no shortage of interesting new problems that could be productively tackled. With the advent of recombinant DNA techniques in the 1970s, anything seemed possible—even the problems of differentiation and pattern formation.

Indeed, by the late 1970s, the subject of morphogenesis seemed itself to be staging a comeback in the biological literature. But it would be difficult to argue that mathematical models played much of a role in this revival (see Chapter 6). Far more significant was the emergence of powerful new techniques for deciphering the genetic program encoded in DNA, coupled with the expectation that it would be possible to read the entire story of biological development from this program.[61] Biologists interested in morphogenesis, even those who had earlier been skeptical of the usefulness of molecular biology for this subject, now had at their fingertips a host of new tools that were not only easy to handle but, more importantly, guaranteed to yield concrete results. Whether or not these would provide what the mathematical biologists regarded as an explanation, they would be recognized and valued by their own colleagues. Where, then, was the incentive for learning an alien (and more demanding) set of techniques?

Even less appealing was the alternative—that is, accepting an invitation to perform experiments whose sole purpose was to support or refute a model on the one hand motivated by unrealistic assumptions, and on the other, generated by tools of a discipline that was not only alien but also somewhat notorious for its colonizing impulses. The modelers were still intruders, still perceived as arrogant and, worse yet, still ill informed about what they saw as the most essential considerations.[62]

From the perspective of biologists, the patterning of body segments in *Drosophila,* one of the most widely studied applications of Turing's model, provides a particularly telling example.[63] Over the last twenty years, experimental analysis of this process has made enormous strides, and it has become one of the great success stories of the new molecular developmental biology. Largely as a result of the work of Christiane Nüsslein-Volhard, Eric Wieschaus, and Ed Lewis (for which they were awarded the Nobel Prize in 1995), we now have a detailed picture of the ordered sequence of events giving rise to segmentation. Rather than reaction and diffusion, pattern formation results from a cascade of gene expression that begins with a specific spatial distribution of mRNA molecules and transcription activators already laid down in the egg. Diffusion plays a role in this process, but it is the progressive activation of a hierarchy of genes that defines the final pattern. As Maini and colleagues write, "Although RD theory provides a very elegant mechanism for segmentation, nature appears to have chosen a much less elegant way of doing it!"[64]

From the vantage of our current understanding, Waddington's reservations are also vindicated, and on all counts. Turing's model may well have proven useful for analyses of two-dimensional pattern formation,[65] but for the analysis of individual development—which, as Waddington pointed out, rarely if ever proceeds from "a featureless sphere"—it has been considerably less so. Addressing itself to the question of how embryogenesis *could* work, it missed the most essential features of how biologists now believe it *does* work.[66] It has become evident that genes cannot be eliminated from the discussion—they are essential to every part of the process. Furthermore, the work of Nüsslein-Volhard and Wieschaus has provided one of the most powerful demonstrations to date that animal form does not emerge *de novo*. They have shown that inhomogeneities in the unfertilized egg cannot be ignored, any more than genes can. The specific spatial patterns of proteins and messenger RNA already in place, preformed as it were, are essential for embryogenesis, for they are what set in motion the sequence of events that lead to the development of the adult form. Waddington's final reservation can also be seen as prophetic: current research has shown that even apparently stable biological patterns are rarely "once for all and merely there" but are, rather, dependent on systems of ongoing regulation. Indeed, the influence of Turing's reaction-diffusion model in mathematical biology seems finally to have peaked, and what is now seen as most promising are models of self-maintaining dynamic processes (or pathways) of just the kind that Waddington sought to include under his category of "stable dynamic equilibria."

Turing's foray into biology was of immense importance for the study of chemical systems, for the development of the mathematics of dynamical systems, even for many problems in physics. But not, it would seem, for developmental biology. Neither he nor the generations of mathematical biologists he inspired succeeded in healing the divide that D'Arcy Thompson had already observed fifty years earlier. Rather than from mathematics, the best explanations of development we have today have come from experimental genetics, and it is to this subject that I now turn. But the story is not yet over. In Chapter 8, I discuss a number of indicators pointing to at least the beginnings of a rapprochement between the cultures of mathematical and biological science and suggesting that the day of mathematical biology may finally have arrived. Might Turing simply have been too soon?

PART TWO

Metaphors: Genes and Developmental Narratives

Far and away the most successful research program of twentieth-century biology has been experimental genetics. We owe to genetics the conceptual framework that has most profoundly shaped the ways in which contemporary biologists think about development. But the science of genetics has undergone dramatic changes over the course of the past century, and, inevitably, so too has its conceptual framework. We can identify three stages in that evolution: (1) classical genetics (roughly 1930–1960); (2) early molecular biology (roughly 1960–1980); and (3) post-recombinant DNA developmental molecular biology (roughly 1980–2000). In each of these stages of its history, genetics introduced its own characteristic way of framing the problem of development.

In the first period, the dominant explanatory framework genetics provided for development rested on the notion of gene action; in the second, on the notions of feedback and genetic programs; and in the third, on the notion of positional information. All of these terms borrow from other domains, carrying meanings and explanatory functions employed in those other domains even when discordant, but they are now put to work in new contexts in ways that exploit both the consonance among these other meanings and the tensions evoked by their various kinds of mutual discordance. Indeed, this very process of readaptation generates new kinds of explanatory force. Successive stages of developmental narratives in genetics need not dispense with earlier constructions; rather, just as in the evolution of species, they can, and often do, embed earlier meanings in new constructions.

Thus, the explanatory force of the terms gene action, feedback, genetic programs, and positional information relies upon and makes use not only of their function in the disparate contexts from which they were borrowed but also of the functions of earlier terms and earlier forms of explanation in genetics that may no longer be explicitly invoked. Just how this works is the question I want to explore in this section: how the forms of explanations of development offered by genetics acquire their force, and how they differ from explanations relying on material and mathematical models (like those featured in Part One). First, however, two caveats concerning nomenclature.

The first caveat concerns the meaning of the word experimental in the practice of genetics, especially in relation to its companion term theoretical. In invoking the modifier experimental, I do not wish to imply that research in laboratory genetics is atheoretical; I merely wish to distinguish it from the kind of work on mathematical models of evolving populations that is sometimes referred to as theoretical genetics (though more commonly as population genetics). Experimental genetics is itself theoretical, even if not in the sense of depending on mathematical models. It is theoretical, first, in its dependence on gene theory (either classical or molecular), that is, on a system of presuppositions about the nature and functions of the gene, and, second, in its analysis of possible relationships which can be brought into consistency (or inconsistency) with the data at hand—analyses that are inevitably required for both the interpretation of experimental data and the design of new experiments.

For this reason, the demarcation I draw between experimental and theoretical (or population) genetics operates primarily between two subdisciplines. It is motivated—in fact, for the purposes of this discussion, mandated—by the virtual silence of population genetics ever since its beginnings in the 1930s on the problem of individual development and by the corollary fact that, in both classical and molecular genetics, efforts to explain development have depended entirely on the results of controlled laboratory experiments performed on so-called model organisms. But here too a caveat is in order: this time about the meaning of the term model. Unlike the experimental work of synthetic biology, experiments in genetics are not performed on artificial organisms but on actual organisms. Indeed, the primary meaning of the term model in experimental biology *is* an organism, an organism that can be taken to represent (that is, stand in for) a class of organisms. A model in this sense is not expected to serve an explanatory function in itself, nor is it a simplified representation of a more complex phenomenon for which we already have explanatory handles. Rather, its primary function is to provide simply a stable target of explanation. The explananda—the phenomena that need to be explained—are the patterns of behavior exhibited by a model organism under well-defined experimental conditions.

With these caveats in mind, we can turn to our principal questions: How are explanations of development constructed in genetics? How are they grounded in experiment? What kinds of conceptual tools are employed in crafting such explanations out of genetic data? And how, once formed, do they persuade (or fail to persuade)? Geneticists do not rely on the tools most familiar to physical scientists. They are well known to be unsympathetic to mathematical models of the kind that Rashevsky, Turing, or their successors developed. They have shown equally little interest in mechanical analyses of developmental processes, either in those that D'Arcy Thompson found so promising or in more recent efforts in biomechanics. And they would surely have disdained explanatory efforts based merely on visual similitude to processes that were not themselves grounded in the material realities of real biological systems (Leduc's, for example).

In this, they have shared the epistemological values of most experimental biologists: an explanation, to be acceptable, ought to tell us how the systems we study do in fact actually work—not how they could hypothetically work. Furthermore, geneticists add to this another criterion that is not always shared by other experimental biologists (and was in fact far from widely accepted among geneticists themselves in the first half of the twentieth century), namely, that a satisfying explanation must acknowledge the critical importance of genes and accord them a correspondingly central role. By this criterion alone, all of the models discussed in Part One are guaranteed to fail; and by the same criterion, genetic explanations of development automatically constitute a class of their own.

But what kind of role might genes play in explanations of development? The answer would seem to depend on what kind of entity the gene is taken to be. What then *was* a gene to geneticists working in the first half of the century? Typically, models work in scientific explanations by choosing elements about which principles governing their behavior are already known. (I think here of Maxwell's mechanical models for the electromagnetic field or of the examples discussed in Chapter 1.)[1] However, the fundamental elements of genetics are like none of these: they are not atoms, molecules, gases, fluids, or solids obeying well-established physical laws; nor are they simple chemicals behaving in accordance with familiar laws of chemical reactions. Of course, Mendel's rules of inheritance do provide reasonably good guides to the sorting of genes in genetic crosses, but these rules are not much help in understanding the behavior of genes in development. Furthermore, the very concept of the gene has mutated radically over the course of its history, and even at a given moment in time it has been subject to a conspicuous multiplicity of meanings, requiring for disambiguation an extensive familiarity with the specific context in which it is invoked.

In short, genes display neither the stability nor the clarity expected of the explanatory elements upon which the physical sciences have come to rely. Thus, an examination of genetic explanations of development needs to begin with the meanings attributed to the gene in the particular temporal contexts in which that concept is employed. Also, and more to the point, it will have to make sense of the role that genes, however they are conceived, are made to play in the construction of satisfying accounts of development.

The core of my argument is that much of the theoretical work involved in constructing explanations of development from genetic data is linguistic—that it depends on productive use of the cognitive tensions generated by multiple meanings, by ambiguity, and, more generally, by the introduction of novel metaphors. If material and mathematical models provided the primary explanatory tool in Part One, and computer simulations play a parallel role for the forms of explanation that will be discussed in Part Three, metaphor—understood in the largest sense of that term—may be said to serve as the principal explanatory tool in this section. That is a strong claim, and it calls for at least a few preliminary remarks on the role of metaphor in science.

Most philosophers, and even many scientists, have long since abandoned the traditional view of scientific language as, ideally at least, literal and univocal, uniquely corresponding to the entities and processes that make up the real world. But the specter this tradition cast on the use of metaphors and other linguistic tropes in science dies hard, and the conviction persists among some that when language is not literal it is therefore less than literal—at best, that metaphoric language offers a provisionally useful heuristic to be dispensed with as soon as possible, and at worst (as both Hobbes and Locke believed), a merely ornamental or downright deceptive intrusion that ought not to be admitted in proper scientific discourse.[2] Yet, as historians and philosophers have increasingly come to appreciate, close observation of scientists at work, either in the present or in the past, reveals that they simply cannot function under such a harsh mandate. The difficulty is obvious: scientific research is typically directed at the elucidation of entities and processes about which no clear understanding exists, and to proceed, scientists must find ways of talking about what they do not know—about that which they as yet have only glimpses, guesses, speculations. To make sense of their day-to-day efforts, they need to invent words, expressions, forms of speech that can indicate or point to phenomena for which they have no literal descriptors. As Mary Hesse reminds us, "The world does not come naturally parcelled up into sets of identical instances for our inspection and description."[3] Making sense of what is not yet known is thus necessarily an ongoing and provisional activity, a groping in the dark; and for this, the imprecision and flexibility of figurative language is indispensable.

Of course, the notion that metaphor (along with other figures of speech) can serve important positive functions in scientific explanations is not new. Max Black pioneered the study of the kind of work that metaphorical statements perform in the 1950s and early 1960s, likening their function to that of models (he called them "analog-models").[4] In his widely influential "interaction view," metaphoric utterances are links not just between two words (such as "society is a sea") but between two "systems of associated commonplaces" which the two words separately evoke. Furthermore, the linkage works not by virtue of any pre-given set of properties in which the terms are already known to resemble one another, but by calling forth similarities, by leading the reader to selectively focus on those properties of each system that make a fit between the two terms possible. The process, Black argued, leads us to see the primary referent (society) in new ways, ways that accord to our various associations with the secondary referent (sea). To a lesser extent, the same process also works in reverse, and inevitably so, adapting our perceptions of the secondary referent (a sea) to our conceptions of the primary referent (society). Such metaphoric utterances can be scientifically productive just because they open up new perspectives on phenomena that are still obscure and ill-defined and about which clarification is achieved only through a process of groping—in other words, on the kinds of phenomena that scientists take as the objects of their investigations.

Since Black's initial formulations, philosophers have produced a large body of work on the subject of metaphor in science, much of it focusing on his claim that referential imprecision can have a positive function in scientific work. Hesse has been one of the principal contributors to this literature, and she goes significantly further than Black. Where Black's arguments presuppose a background of literal meanings, Hesse argues (along with Hans-Georg Gadamer) that metaphor is in fact primary to literal meaning.[5] In this view, language (both scientific and "ordinary") is metaphoric "through and through," and literal meanings are seen as emerging only as the end product of a long process of creatively deploying forms of discourse that are themselves (and unavoidably so) imprecise, protean, and ever changing.[6] This, then, is the sense in which Hesse writes of scientific theories in general that they "are models or narratives, initially freely imagined stories about the natural world, within a particular set of categories and presuppositions which depend on a relation of *analogy* with the real world as revealed by our perceptions" (p. 51). Science remains progressive in Hesse's view, first, by virtue of its characteristic forms of "test-and-feedback," grounded in "predicting and controlling empirical events by means of experiment and theory-construction" (p. 52), and, second, as a consequence of the ever-increasing availability of more sophisticated means for prediction and control. Progress here need not be measured in terms of an ideal of universal truth; indeed, it need not imply any such ideal. For most scientists, pragmatic success in approximately describing and verifying particular local phenomena is quite good enough.

I am sympathetic to Hesse's brand of "moderate realism," but I think something can be said for global narratives that guide scientific research in ways that do not directly depend on either test-and-feedback or verification—for narratives that make productive use of the imprecision of metaphor and other linguistic tropes not so much as a way of guiding us toward a more precise and literal description of phenomena but rather as a way of providing explanatory satisfaction where it is not otherwise available. This, at least, is the case I want to make for genetic narratives of development. Just because we have no access to rules or laws describing the role of genes in development, gene-based accounts of these processes need to make use of the associations generated by metaphor, by ambiguity, and by the dynamic interplay among the different meanings a given term may connote. For these purposes, even the absence of a clear and univocal meaning of the very concept of the gene—that is, of the basic explanatory unit—can be a positive resource for drawing different experimental systems and different research programs into a coherent scientific agenda.

This can be seen with particular clarity in the ways in which classical geneticists sought to make sense of development through the notion of gene action. Later, with the advent of molecular genetics, the gene acquired a concrete referent, a specific region of a molecule of double-stranded DNA, and with that advance, the earlier figure of gene action, depending as it had on the very uncertainty of the definition of the classical gene, could no longer satisfy. Geneticists required a new kind of narrative for thinking about development; and to fill the gap left by the demise of gene action, a correspondingly new figure of speech was introduced: the genetic program. In Chapter 4, I extend my own and others' discussions of the history of these two linguistic structures and examine the detailed workings of metaphor and multi-vocality in their respective uses.[7] The subject of Chapter 5 falls chronologically between gene action and genetic programs: it centers on the plasticity of the notion of feedback, and the equally plastic use that was made of one of the rare mathematical models (if not the only such model) to have entered the mainstream literature on developmental genetics during this middle period, namely, Max Delbrück's model of cross-reacting metabolic pathways. Finally, in Chapter 6, in an attempt to do justice to the findings of researchers in developmental genetics over the last two decades, I turn to a new developmental trope that has recently acquired considerable popularity in this literature, namely, positional information.

CHAPTER FOUR

Genes, Gene Action, and Genetic Programs

The meaning of speech thus keeps changing in the act of groping for words without our ever being focally aware of the change, and our groping invests words in this manner with a fund of unspecifiable connotations. Languages are the product of man's groping for words in the process of making new conceptual decisions, to be conveyed by words.

Different languages . . . sustain alternative conceptual frameworks, interpreting all things that can be talked about in terms of somewhat different allegedly recurrent features. The confident use of the nouns, verbs, adjectives and adverbs, invented and endowed with meaning by a particular sequence of groping generations, expresses their particular theory of the nature of things.

Michael Polanyi, *Personal Knowledge* (1958)

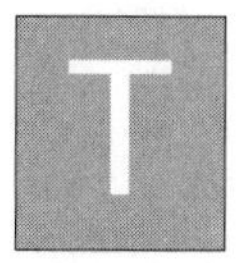

he official history of the gene does not begin until Wilhelm Johannsen coined the term in 1909, only three years after William Bateson had proposed the name genetics for the new studies of hereditary phenomena that had been launched by the rediscovery of Mendel's laws in 1900. But while the introduction of new terms may mark starting points for official histories, in relation to conceptual histories it would be better to think of them as providing signposts, often marking turning points in scientific and intellectual history. And so it proved here. Before the birth of genetics, and before the advent of the term gene, a science of hereditary phenomena already existed under the apparently unproblematic designation "heredity." But at the end of the nineteenth century, heredity was a far more inclusive term than it is today. It encompassed both the study of the conservation of traits across generations and their intragenerational emergence (or transformation) over the course of an organism's development from a fertilized egg. The founding of a new discipline called genetics marked a rupture in this unity, and what had previously been seen as two aspects of a single subject (transmission and development) now came to be seen as distinct concerns.

Thus, two separate disciplines emerged in the early decades of the twentieth century, with (at least initially) two distinct agendas: geneticists studied transmission, while embryologists studied development. In his 1926 book *The Theory of the Gene,* T. H. Morgan described the relation between the two disciplines as follows:

> Between the characters, that furnish the data for the theory and the postulated genes, to which the characters are referred, lies the whole field of embryonic development. The theory of the gene, as here formulated, states nothing with respect to the way in which the genes are connected with the end-product or character. The absence of information relating to this interval does not mean that the process of embryonic development is not of interest for genetics . . . but the fact remains that the sorting out of the characters in successive generations can be explained at present without reference to the way in which the gene affects the developmental process.[1]

Elsewhere in the same year, Morgan cautioned that "the confusion that is met with sometimes in the literature has resulted from a failure to keep apart the phenomenon of heredity, that deals with the transmission of the hereditary units, and the phenomena of embryonic development that take place almost exclusively by changes in the cytoplasm."[2]

But to what extent can the study of transmission in fact be kept apart from that of development? On one level, such a division of labor seemed to make perfect sense, and its meaning was clear. The day-to-day practices of researchers in the two disciplines, their questions, and even the organisms on which they worked were completely different. And for a new discipline seeking to establish itself, it also made good political sense.[3] But on a more conceptual level, it is difficult to see how that division could have been meant as anything other than provisional, as anything more than a temporary and necessarily unstable *modus operandi.* For what would be the point of tracking the transmission of hereditary factors unless these factors can be assumed to be implicated in the formation of the traits or characters that distinguish an organism, that is, in its development? Morgan himself was by this point in his career fully convinced that, however important the cytoplasm might be for development, its influence was at best an intermediate one; in the long run, it was to the genes that one had to look for ultimate causes. As he had written just two years before, "It is clear that whatever the cytoplasm contributes to development is almost entirely under the influence of the genes carried by the chromosome, and therefore may in a sense be said to be indifferent."[4]

In retrospect, the proclamation of a division of labor between the two disciplines might even appear somewhat disingenuous; certainly, not many embryologists were reassured. Reading the claims geneticists had already begun to make for the causal primacy of genes (for example, that "genes are the primary internal agents controlling development"; that everything else in the cell is "only a by-product of the action of the gene material"),[5] they began to worry that geneticists would invade their own discipline. As indeed they soon attempted to do. In the 1930s, geneticists began to tackle the problem of embryonic development directly—in effect, trying to heal the rift between the disciplines to which they themselves had earlier contributed. But in order to do so, they needed first to reformulate the problems of embryology so as to make them more compatible with their own perspectives. And even before that, they needed a way of thinking and talking about genes that would enable these entities to account both for the geneticist's observation of intergenerational constancy (transmission) and for the embryologist's observations of intragenerational transformation (development). What I have called the discourse of gene action provided just the solution they needed, and it built not only on the ways of talking about genes which they had forged in the preceding decade but also on the prehistory of the very notion of a gene.[6]

Indeed, a foreshadowing of the struggle of geneticists (one might even say, of the century-long struggle) to reconcile the two disparate demands the gene was called upon to satisfy—that is, accounting both for intergenerational transmission and for intragenerational development—can already be glimpsed in Hugo de Vries's early writings, eleven years before the rediscovery of Mendel's laws in 1900, as he groped for an adequate description of his pangens. On the one hand, de Vries invoked the established elements of physics and chemistry to justify his own hypothesis of particulate elements: "Just as physics and chemistry are based on molecules and atoms, even so," he wrote, "the biological sciences must penetrate to these units in order to explain by their combinations the phenomena of the living world."[7] On the other hand, if such a unit were, as he put it, to "impress its character upon the cell" (p. 194), to either "represent" the properties of an adult organism or cause their coming into being, it must obviously be something larger and more complex than a chemical molecule. "These minute granules," he concluded, "are more correctly to be compared with the smallest known organism" (p. 4).

When Wilhelm Johannsen felt called upon to coin a new word twenty years later, it was in large part in order to liberate the study of genetics from the taint of preformationism carried by the image of a hereditary unit *qua* organism. He proposed the term gene in the hope that a new word would be entirely free of such hypotheses. In 1911 the Dutch geneticist Avend Hagedoorn echoed Johannsen's sentiments: "It is inadmissible," wrote

Hagedoorn, "to try to explain the facts of evolution and inheritance by the behaviour of living particles which have been invented simply to admit of this explanation."[8] In hindsight, long after genes had come to supplant both de Vries's pangens and Weismann's determinants, H. J. Muller came to think of these earlier concepts as betraying a subconscious adherence to "the ancient lore of animism."[9] Yet even with a new term, de Vries's ambi-valent image endured.

As critics of the new discipline frequently observed in the early decades of genetics, the fundamental unit of heredity persisted—even if only tacitly—as an entity that was at one and the same time both atom and organism.[10] Such enduring tension in the structure of the classical gene can be seen in Leonard Thompson Troland's early efforts to reconcile the duality of function it was required to serve;[11] it can be seen in Muller's own writings; and, perhaps above all, it can be seen in the notion of gene action. Furthermore, it was just this hybridity in the concept of the gene that lent cogency to the discourse of gene action, and also that provided so apparently natural a way to reframe the problem of embryonic development.

Alfred H. Sturtevant provides a particularly clear example of just such reframing. In his classic paper on the developmental effects of genes, presented at the 1932 International Congress of Genetics, he explained: "One of the central problems of biology is that of differentiation—how does an egg develop into a complex many-celled organism? That is, of course, the traditional major problem of embryology; but it also appears in genetics in the form of the question, 'How do genes produce their effects?'" (p. 304). Furthermore, he continued, "in most cases there is a chain of reactions between the direct activity of a gene and the end-product that the geneticist deals with as a character"; genetic experiments can therefore approach the problem through the "analysis of certain chains of reactions into their individual links" (p. 307). As Michael Polanyi writes, "The confident use of the nouns, verbs, adjectives and adverbs, invented and endowed with meaning by a particular sequence of groping generations, expresses their particular theory of the nature of things." Sturtevant, in his reformulation of the classic problem of embryology, confidently employs the language of his own generation of scientists groping their way to an understanding of genetics, a language that expresses their particular view of the relation between genes and the "end-products" they "produce," and between sets of genes and the ensemble of characters that make up an organism. Gene action is a shorthand expression for this way of thinking: it represents development as proceeding along chains of reaction that start with fertilization (the event that triggers the onset of gene action) and culminate in the production of an organism seen as an effective summation of the end products of the activity of all its genes.

The term was routinely employed by geneticists throughout the 1930s, 1940s, and 1950s—and surviving even into the 1960s—as a way of both describing and explaining the role of genes in development.[12] But what does it mean to speak of gene action? The first part, gene, remains quite ill-defined during this entire period, but the meaning of action, while broad enough, seems fairly clear and not conspicuously fraught with ambiguity. We know what it means to act, and we have a more or less clear idea of the kinds of entities that are capable of acting and, hence, to which notions of agency can be attached. In the first instance it is of course persons we think of as acting, or animals, or perhaps even organisms in general. But also, by an extension that is of interest in itself—and here a certain ambiguity might indeed be said to enter—chemical enzymes are by this time in history entities that can also be said to act.[13] This latent ambiguity needs to be kept in mind as we turn to the matter of primary interest, namely, the conjoining of "action" with "gene."

What is a gene, and what kind of action might it be capable of? This, of course, no one working or writing in the 1930s could say. At the time, the gene was still a hypothetical entity—an entity that had originally been invoked in a reach toward an explanation of living phenomena—or, put more strongly, of the "riddle of life." It is a term that was introduced to fill a lexical gap, and for the moment we might, just as early generations of geneticists were obliged to do, suspend the question of just what a gene is. But then we immediately find there is also another question, namely, what does a gene do? Here is another lexical gap—we have no word to describe the function of a gene. Indeed, it is to supply the missing term that the conjunction of "gene" with "action" is called forth.

How this conjunction works may be partially illuminated by Aristotle's classic analysis of an analogous problem in his *Poetics,* namely, of how one might speak of the life-giving powers of the sun (that is, of the relation between the sun and its rays) and in particular the use of the expression "the sun sowing" its flame. Aristotle writes: "It may be that some of the terms thus related have no special name of their own . . . Thus to cast forth seed-corn is called 'sowing'; but to cast forth its flame, as said of the sun, has no special name. This nameless act (B), however, stands in just the same relation to its object, sunlight (A), as sowing (D) to the seed-corn (C). Hence the expression in the poet, 'sowing around a god-created *flame.*'"[14] The parallel with the expression gene action is clear as far as it goes: we do not know what it is that the gene does, but a name for that function is provided by likening the gene to that which we do know to be capable of action (that is, an agent). But the parallel is only partial, for, whereas we know what the sun is, we do not know what a gene is. In the case of gene action, we thus find not one but two lexical gaps—first, a word and, second, a denotation for the subject of that missing word. Our figure of speech needs therefore to do more work than the poetic expression "the sun sowing." It must serve a dual function—first, to fill the gap created by the absence of a word for what the gene *does,* and, second, to fill the gap created by the absence of a clear referent for the gene itself.

How, we need to ask, does it meet the second need? My answer is this: the gene is not merely likened to the person (or animal) who acts (as the sun is likened to the farmer who sows), but—in what might be described as a doubling of the metaphoric movement—it actually acquires one part of its definition by that comparison. In other words, the gene is defined as an entity embodying the capacity to act (in whatever ways are required) within its own being (recall, for example, de Vries's "smallest known organism"). At the same time, however, because of the latent ambiguity that had come to inhere in the word action as a result of its incorporation into the language of chemistry, the phrase also likens the gene to a chemical object (especially an enzyme) and, accordingly, constitutes the gene by that comparison as well. In this way, the fundamental uncertainty in the nature of the gene simultaneously invites and is reinforced by its association with the word action—a double movement that can endow the resultant expression with even greater rhetorical efficacy.[15]

The Need for an Essential "Ambi-Valence" in the Concept of the Gene

I have already noted that, for purely practical reasons, geneticists in the 1930s had little choice but to leave a specification of the gene's particular characteristics in abeyance. Yet the fact is that at least two aspects (or functions) of the gene—corresponding to the two associations we have just seen—were built into the ways in which the term had been used ever since it was introduced in 1909. Like Mendel's own "elemente" (1859), it was to account for the patterns of inheritance in genetic crosses, but like

Weismann's determinants (1885) and de Vries's pangens (1889), it was also (as Weismann's term in fact suggests) to account for the determination of an organism's properties. Even in the absence of concrete evidence concerning the nature of the gene, one would have had to ask: What kind of an entity might be imagined as serving such radically different functions?[16] And part of what made it so difficult to imagine how such an entity might be physically constituted was the essential duality (or ambivalence) it would need to contain within itself. For how else, to put the question as Arnold Ravin has posed it, "Would it be possible to reconcile, within the context of a material hereditary factor, the properties of transmission and potency in developmental determination?"[17]

From its very beginning then, the gene was already something of a monster—not quite a metaphor, or at least not in any strict sense of the term, but a neologism that has the potential of even greater force for it builds on the work of two or more metaphors that are between or among them not only disjoint but in active tension, and conjoins these into a new and apparently seamless unity.[18] Where a metaphor says "this is that," inviting us to see both this and that in new ways, here we have a linguistic construction that amalgamates this and that. It melds into a single form two entities with the disparate properties of atom and organism, and contains the incoherence of such a melding under the protective wrap of a new word. In effect, it offers a resolution of the riddle of life by invoking an entity that is a riddle in and of itself.[19]

It may be that such evidence of what might be called primitive incoherence will tempt the reader to conclude that early generations of geneticists were simply confused or self-deceiving, but such an inference would be a serious mistake. In the first place, it would be manifestly presumptuous, and inexcusable for that reason alone; but more importantly, it would be unrealistic: it would be to discount the productive value of such incoherence. Without it, geneticists would have been left without a framework with which to proceed. Insofar as the science of genetics was committed to a causal narrative of development, there needed to be a place, a thing, a word to which causal force could be attached, and in the absence of any foreseeable route to a clarification of what that thing might be, it was functionally important to have a word that could contain or black-box its uncertainty and to keep its internal incoherence, as it were, under wraps. Not only did it permit researchers to get on with their work—resolving the problems they *were* able to address, without having to worry about those they could not—but also it provided them with an explanatory framework, albeit a provisional one, with which they could make sense of the progress they were making in their day-to-day research, both to themselves and to others. That it also served other uses—for example, in the struggle to establish genetics as a discipline and even in international and gender politics—only means that it was productive in more than one sense.[20]

But such strategies of containment can hardly be productive in all ways, and, needless to say, this one was not. The unmasking of incoherence, bringing it into the open, is well known to be scientifically productive as well: it spurs researchers into asking new questions and seeking new formulations. Thus, the containment of this particular incoherence inevitably had its costs. While, on the one hand, it helped to promote a consensus within the discipline that was manifestly beneficial, it also served to insulate practitioners from the stimulus of perceived contradiction and, equally, from the demands of critical challenge. One episode provides a case in point that is not only illustrative but of particular historical importance.

In 1934, having turned his attention once again to the subject of embryology, T. H. Morgan had second thoughts about gene action, and, from his new perspective, offered an incisive critique of the notion and an alternative proposal to put in its place. He wrote:

> The implication in most genetic interpretation is that all the genes are acting all the time in the same way. This would leave unexplained why some cells of the embryo develop in one way, some in another, if the genes are the only agents in the results. An alternative view would be to assume that different batteries of genes come into action as development proceeds . . . The idea that different sets of genes come into action at different times . . . [requires that] some reason be given for the time relation of their unfolding. The following suggestion may meet the objections. It is known that the protoplasm of different parts of the egg is somewhat different . . . and the initial differences may be supposed to affect the activity of the genes. The genes will then in turn affect the protoplasm . . . In this way we can picture the gradual elaboration and differentiation of the various regions of the embryo.[21]

Eric Davidson has credited these penetrating remarks as offering "the first explicit statement of the theory that differentiation could be caused by *variation in the activity of genes in different cell types.*"[22] In hindsight, they even seem obvious. Yet both the critique and the proposed alternative went largely unheeded for more than two decades.[23] Thus, the salient question is: Why would an observation that (at least to modern ears) seems so transparently sound, one that had been made by the preeminent leader of the field, have for so long had so little impact? Why, in short, did not Morgan's critique have more bite?

But if the question is evident, the answer should be equally so: entities that were tacitly endowed with animate agency, as genes were, *might* have the foresight that would enable them to direct "some cells of the embryo in one way, some in another." Such an expectation need not ever be made explicit—it is enough for it to reside in the background, never spoken yet always available. In this way the very uncertainty in the nature of the basic hereditary unit becomes a resource. In other words, that Morgan's critique had no bite can be understood as a consequence of the fact that the notion of gene action was too labile, too dynamic; the slide between the different conceptions of the gene that it evoked was so easy that it had become automatic, and the effect was to leave the notion of gene action without anything sufficiently solid for such a critique to bite into. One might say the downside of that construct was a consequence of its being *too* productive.

Of course, the very fact that such a combination of properties was so mysterious was certainly provocative, and it did impel at least a number of researchers to search for a concrete and recognizable entity that could fit the bill. To some extent, therefore, the ambi-valence of the gene was also productive in the traditional way, as leading to a univocal and more literal definition—as a spur to its own demise. Yet when the genetic material was finally identified as made up of DNA and that quest appeared to have been realized, something was lost. There was no doubting the reality of DNA; and the discovery of the structure of the double helix, together with that of a correspondence between sequences of nucleotides and sequences of amino acids, did (or so it seemed at the time) cleanse the concept of the gene of all residual ambiguity. But for the purposes of explaining development, these findings also cleansed it of its force. Once the gene could no longer freely oscillate between atom and organism, it could no longer serve so readily both as the fundamental unit of heredity and, at the same time, as the pilot of life's developmental journey. Thus, for a satisfying global narrative of development—especially, for an account of how genes might direct "some cells of the embryo in one way, some in another"—a new linguistic construct would be needed, one that would accommodate the new information, even make room for Morgan's discerning comments, and yet rival gene action in its ambi-valence, its flexibility, and its power.

New Genes, New Metaphors: DNA and Genetic Programs

Histories of the molecular revolution in biology abound, and a good deal has been written on the particular subject of its metaphors of information and programs.[24] My focus here, however, is more narrow: it is on the explanatory force of the concept of the genetic program. This concept, explicitly borrowing from computer science, was first introduced by François Jacob and Jacques Monod in 1961 in an attempt to extend their success in analyzing a mechanism for the regulation of enzyme synthesis in *E. coli* to a more general description of the role of genes in embryonic development—that is, to a resolution of the problem that had been noted by Morgan a quarter of a century earlier.[25]

Concluding their first published report of what is now commonly referred to as the operon model, they wrote: "The discovery of regulator and operator genes . . . reveals that the genome contains not only a series of blue-prints, but a coordinated program of protein synthesis and the means of controlling its execution."[26] And, in a more general discussion that same year, they claimed to have shown from their studies of biochemical regulation in bacteria that cell differentiation in higher organisms "does not constitute a 'paradox,' as it appeared to do for many years, to both embryologists and geneticists" (p. 397): by implication at least, this "paradox" had been resolved by Jacob and Monod's own demonstration of the existence, within the genome, of a "coordinated program of protein synthesis and the means of controlling its execution"—in other words, by the existence of a genetic program.

Elsewhere, I have tracked the rise of the concept of a genetic program in contemporary biology, both noting its lack of definition (what in fact *is* a genetic program?) and arguing against its explanatory adequacy for thinking about development.[27] Here, however, I want to take the currency which the genetic program has acquired in gene-based narratives of development as a given and focus not on its putative shortcomings but rather on what many (perhaps even most) biologists see as its strengths. How can we better understand the fact that this concept, ill-defined as it is, has so manifestly offered long-term explanatory satisfaction, and to so many? How are we to reconcile the numerous criticisms which have been put forth—all of which in any case derive from concerns with which biologists are already abundantly familiar—with the observation that, for reasons of their own, so many researchers in the field do not see these as being the slightest bit problematic?

The concept of a genetic program, I contend, provides the essential ingredient required for a global narrative relating genes that can be variably activated to the development of fully formed organisms. It does so, first, by virtue of an essential ambiguity in the locus of agency and, second and more immediately obvious, by its metaphoric invocation of a computer program. Indeed, the computer metaphor by itself might seem to provide adequate insulation from the ancient taint of animism that Muller had diagnosed. But what I refer to as the essential ambiguity was equally important, and it can be seen even without reference to the computer metaphor: the term program has meaning in and of itself, acquired long before the advent of computers. It suggests a plan of procedure, a schedule, or even a set of instructions. But what exactly is a genetic program? Indeed, the very same question can equally well be asked of a host of related (and closely linked) terms: genetic control, genetic regulation, genetic switches, and even genetic activation. With all these terms, the central problem (and correspondingly central ambiguity) becomes evident as soon as we ask about the grammatical case to which the modifier genetic refers: Are genes to be understood as the subject or as the object of the genetic program? Are they the controllers or the controlled, the regulators or the regulated, the switches or the entities to be turned on and off, the activators or the activated? If they are controllers, regulators, and activators, even without the help of a computer metaphor it is a short step to think of them as issuing instructions. But if they need to be controlled, regulated, and acted upon, then, presumably, instructions would have to come from elsewhere. Furthermore, we might note yet another linguistic slippage lurking here, for "acted upon" is not quite the same thing as "activated." Even if genes need to be activated, do they, once activated, then become agents capable of action? None of these uncertainties can be regarded as either incidental or separable from the others; indeed, one played off the other. Moreover, by doing so, they collectively constituted an essential resource from which the term genetic program could draw both its meaning and its force.

The need for a new figure of speech had arisen directly from the recognition that, after all, just as Morgan had conjectured, genes do not act all the time—they need to be activated and inactivated, turned on and off. Although the first compelling evidence for this conclusion came from studies of enzyme synthesis in bacteria, the extrapolation to higher organisms was made at once. In fact, one might say that these bacterial studies licensed acceptance of the notion of variable gene activity in higher organisms—in organisms that, unlike bacteria, undergo cell differentiation and development and therefore require some such notion. When talk of gene action shifted to talk of gene activation, the immediate reference in the new expression was to genes as the *object* of activation, and not as its subject. Genes, we learned, need to be regulated, controlled, switched on and off, by something else—presumably, by non-genic entities. Should one then conclude that Morgan's suspicion had been right, that genes are not the only agents involved?

The invocation of agents in Morgan's phrasing of the question is critical. To geneticists working in the 1950s and 1960s, at least two considerations argued against such a conclusion, and both derive their weight from the persisting inclination to frame this question in terms of agency and agents. So framed, one is inevitably led to ask: What else, besides the genes, could qualify? Might, for example, one find agency located in the cytoplasm? At that time, the notion of cytoplasmic agents clearly implied cytoplasmic genes, but this possibility had long been disparaged by most classical geneticists, and (with the exception of mitochondrial genes) effectively dismissed by molecular geneticists.[28] Where else, then, might one look? As it happened, an entirely new option had now appeared on the scene. Indeed, the literature of the time shows that genes acquired (or regained) the status of subjects capable of activating, regulating, and controlling as part and parcel of the discovery of gene regulation.

The crucial new option appeared with the identification of particular sites on the DNA (and hence genetic elements) whose presence proved necessary for the regulation of those genes directly involved in enzyme synthesis. To distinguish what were now clearly two different kinds of genetic elements, Jacob and Monod called the latter "structural genes" and the former "regulatory genes." By their very name, regulatory genes could be seen as stepping into the breach, as supplying the missing agency. Structural genes may need to be regulated, but regulatory genes were there to do the job.[29] As the authors of the new labels subsequently explained: "The purely structural (one gene-one enzyme) theory does not consider the problem of gene expression. The discovery of a new class of genetic elements, the regulator genes, which control the rate of synthesis of proteins, the structure of which is governed by other genes, does not contradict the classical concept, but it does greatly widen the scope and interpretative value of genetic theory."[30]

It was of course understood that, in order to do their job and turn the structural genes on and off, regulatory genes themselves needed to be activated, but this fact seemed to present no impedance whatsoever to the new construction. Part of the reason for this is already suggested by the legacy of so long a tradition of agentic discourse in genetics, and by the persistence of ingrained habits of thinking and talking that maintained the capacity to act, to control, or to govern as an inherent property of the gene, even after the gene had been recognized as no more than a chemical molecule, and a relatively inert one at that. To be sure, this legacy had been weakened by the findings of the new molecular biology, and in just the ways I have already indicated. Yet even so, it might at the same time be said to have found new sources of strength in the molecular reformulation, particularly in its deployment of metaphors borrowed from computer science. Indeed, one might say that the agency that genes had lost in their evolution from organisms to DNA was more than compensated for by their newly acquired efficacy as information and program. The role of genes in development could now be distinguished (set both above and apart) from that of other factors in the cell by the "information" they encoded.

At least for the purposes of providing a global narrative, I want to suggest that much of the success of the molecular reformulation rested on new sources of ambiguity that came as part and parcel of the introduction of computer metaphors into genetics. In the ordinary sense of the term, the representation of the specificity of nucleotide sequences as information can scarcely be faulted—in fact, the observation of the extraordinarily high degree of specificity such sequences could carry had lent crucial support to the identification of the genetic material as DNA. And of course, once that identification had been established, the same observation of specificity could be invoked (and so it often was) in reverse, that is, as confirmation of the special status of the genes. But there is another point that needs to be made. By this point in history, the term information had also acquired a more technical meaning, and it was all but inevitable that its use in the biological context would concurrently endow the DNA with some of the special powers that had accrued to the term from its currency in the new science of communication.[31]

The concept of the genetic program built on both the new and the older kinds of ambiguity, and it did so in ways that allowed each to reinforce the other. On the one hand, this expression inherited the ambiguity of grammatical case that we have seen in the related expressions "genetic regulation" and "genetic control." On the other hand, it helped to stabilize the meaning of these other terms by tacitly invoking the metaphor of a computer program. Computer programs at that time had a very specific meaning, namely, a set of instructions encoded in a linear sequence of bits. And even though no one would literally identify a molecule of DNA with a magnetic tape, anymore than they would literally identify the cell with a computer, the analogy was so conspicuous that it could hardly have been missed. Just as in computer programs, so too in DNA, the information that is carried is encoded in digital form, and as in the one so too in the other, that information can be thought of as constituting the intelligence or brain of the machine it directs.

To see how this patently metaphorical use of "program" actually functioned, one need only look to Jacob's own recounting of the history of heredity, written just a few years after the term had first been introduced into genetics. Here, Jacob describes the organism as "the realization of a programme prescribed by its heredity," and, a few pages later, notes that "the programme is a model borrowed from electronic computers. It equates the genetic material of an egg with the magnetic tape of a computer."[32] Quoting Claude Bernard's observation from 1878 that the study of vital phenomena reveals the presence of "a pre-established design," that "some invisible guide seems to direct [the organization and growth of the individual organism] along the path it follows, leading it to the place which it occupies,"[33] Jacob writes, "Not a word of these lines needs to be changed today: they contain nothing which modern biology cannot endorse. However, when heredity is described as a coded programme in a sequence of chemical radicals, the paradox disappears."[34]

Furthermore, to justify this claim, to explain how a description of heredity as a coded program can account for the organized and apparently purposive development of an organism, Jacob refers to the familiar characterization of teleology as a "mistress" whom biologists "could not do without, but did not care to be seen with in public," and he writes, "The concept of programme has made an honest woman of teleology" (pp. 8–9).

These remarks of Jacob's have been quoted many times, and my primary reason for including them here is simply to illustrate the sheer compass (in both magnitude and range) of the force that the concept of a genetic program could be expected to support—particularly in relation to the earlier concept of gene action. No classical geneticist would have so openly claimed for gene action the powers, or intelligence, of an invisible guide capable of leading the organism along its developmental trajectory. Indeed, the very notion of a final cause operating in development, the idea that the passage from zygote to adult was an inherently goal-oriented process, would have been anathema to earlier generations of geneticists. But now, in Jacob's construal of the genetic program, that concept could not only accommodate the recent discoveries of gene regulation but also provide an apparently natural bridge to the long tradition of teleology in biological thought that had been so insistently discredited in the preceding decades.

It may well be that subsequent developments in biological research have created serious difficulties for Jacob's simple reading of the genetic program—especially for his reading of its role in development—and that, in so doing, they have visibly undermined the narrative appeal of that notion.[35] What is at issue here, however, is a different matter altogether: it is the explanatory satisfaction this concept so manifestly provided at the time it was introduced (and to a somewhat lesser extent, continues to provide even in the face of recent challenges) that we want now to better understand. I turn, therefore, to Jacob's invocation of purpose.

To the extent that the instructions encoded on the magnetic tape of a computer can be said to express purpose, it is immediately obvious that it is the purpose of the programmer that is so expressed. In fact, even the term instructions derives its meaning from the intent of the human agent writing the software: the message encoded in the program carries instructions issued by the programmer. In this sense, the Turing metaphor quite conspicuously fails to serve the needs of twentieth-century biologists. If purpose is to be made respectable enough to be explicitly invoked in contemporary explanations of biological processes, it must first be cleansed of any such readily recognizable reference to external intention, whether human or divine. Thus, something more than the image of a Turing tape was required. And, as we well know, this something more did not come from computers per se but from war-related work on self-steering mechanisms.

In 1943 Norbert Wiener and Julian Bigelow, drawing on their own wartime research on self-guiding, goal-seeking anti-aircraft devices, published a paper with Arturo Rosenbleuth entitled "Behavior, Purpose, and Teleology" in which they explicitly equate teleology with negative feedback. A few years later came Wiener's *Cybernetics* (1948) and *The Human Use of Human Beings* (1950)—both of which were immediately and widely translated and helped enormously to popularize his ideas in the culture at large. As in the earlier paper, goal-oriented behavior was the central concern of these popular and semi-popular writings, and so it remained for those of his followers who worked on "self-organizing systems" (sometimes referred to as second-order cybernetics) in the decade to come. In fact, the same concern dominated virtually all of the work on complex systems during this period. And in all these endeavors, the analogy between organism and machine was routinely invoked by Wiener and others, in the Macy Conferences of the late 1940s and early 1950s, by the many other workers in the various fields of systems analysis, and in the conferences on self-organizing systems held in the late 1950s and early 1960s.[36] Indeed, it was precisely because of their apparently goal-directed behavior that organisms provided so apt a model for the new kinds of structures (both mechanical devices and human organizations) that this new breed of scientist and engineer was attempting to build.

Not surprisingly, Jacob makes good use of these precedents. Drawing directly on Wiener, he writes:

> This isomorphism of entropy and information establishes a link between the two forms of power: the power to do and the power to direct what is done. In an organized system, whether living or not, the exchanges, not only of matter and energy, but also of information, unite the components. Information, an abstract entity, becomes the point of junction of the different types of order. It is at one and the same time what is measured, what is transmitted and what is transformed. Every interaction between the members of an organization can accordingly be considered as a problem of communication. This applies just as much to a human society as to a living organism or an automatic device. In each of these objects, cybernetics finds a model that can be applied to the others: a society, because language constitutes a typical system of interaction between elements of an integrated whole; an organism because homeostasis provides an example of all the phenomena working against the general trend towards disorder; an automatic device, because the way its circuits are geared defines the requirements of integration. In the end, any organized system can be analysed by means of two concepts: message and feedback regulation.[37]

In hindsight, it is easy to see that the cybernetic revolution—at least as envisioned by Wiener and his followers—proved to be short-lived. In fact, when Jacob was writing his book, the cybernetic vision had already begun to fade in its home disciplines, at least in part because of the ever more conspicuous gap between its claims and its actual achievements. Furthermore, there remained built into that vision an essential ambiguity of its own, and while this ambiguity was at first an asset, eventually it came to be seen as a liability. Despite the undeniable sophistication of feedback devices and the complex systems of organization they inspired, these structures could only be said to be self-steering and self-organizing in the most limited and patently metaphorical sense. Even while embodying new principles of self-regulation, their obvious function was to extend the range of human control. Nevertheless, by 1970 the language of Wiener's vision (together with its promise) had already been so fully assimilated into the popular imagination that, whatever its technical status, Jacob could still usefully build on Wiener's rhetoric to construct a compelling picture of his own.

What was scarcely noticed, however—in either Jacob's own assimilation of the cybernetic vision or the popular version—was that Wiener's language of purposive, goal-directed, and self-steering machines had been silently conjoined with another language, namely, that of Turing's computational vision, and the dependence of Jacob's argument on this apparently seamless conjunction was vital. On the one hand, and as Jacob himself had explicitly stated, the metaphor of a genetic program drew directly from the computers that were being built on the model of Turing's original thinking machine. But the idea of a purposive machine came from the self-steering devices that had inspired Wiener's cybernetics. These were not only different kinds of machines, built on distinctively different principles, but also they were designed to achieve manifestly different ends: one sought computational power while the other sought the stabilizing effects of self-regulation. Nevertheless, the particular claim that the sequence of the DNA could serve as Bernard's invisible guide clearly required that they be joined together.

Of course, Jacob was by no means alone in taking such an amalgamation for granted. The very intermingling of words like program, information, message, feedback, purpose, and self-organization gave the semblance of self-evident truth to the assumption that, somehow, the differences between these two disparate developments would resolve themselves spontaneously. Yet despite their persistent conjoining in the popular imagination, despite Wiener's own hopes, and despite even von Neumann's efforts at integration, conspicuous differences between the two persisted at least until well into the 1980s.[38]

Recapitulation

Let us try to see what all this has gotten us. We have seen a number of different sources of ambiguity from which the concept of a genetic program was able to draw, and it might be useful to list them. The first and perhaps primary or essential ambiguity derives from the same lack of specification of grammatical case of the modifier "genetic" that already haunted the terms genetic regulation, genetic control, and genetic switch. The second is rooted in the all but inescapable temptation presenting itself at that particular moment in history to liken the digital embodiment of genetic information in sequences of nucleotides with the digital encoding of the literal messages that had formed the subject of information theory, and in the chronic slippage that came as an inevitable consequence of the use of the same word "information" in these two disparate contexts. A third source of ambiguity comes from the realization of the organisms–machine metaphor envisioned by cybernetics; and the last, from the tacit conjoining of the telos of self-steering and self-organizing automata with the intentionality expressly encoded in the digital program of a thinking machine.

The key feature for each of these figures of speech is precisely the referential uncertainty it gives rise to and, because of that uncertainty, the ease with which it invites oscillation between one form and the other. But taken together, we can see another and equally important effect—namely, in their remarkable capacity for lending mutual strength to one another, and in the process lending this collective of figures an air of literality. What then is a genetic program? It is a plan or procedure for turning genes on and off at the right time and in the right place; and, just as instructions are encoded in a computer program, here too we can expect to find the instructions guiding the development of the organism encoded in the digital sequence of nucleotide bases—the trick is only in learning the proper method of decryption. But lest one might think of this program as embodying the intentionality of a programmer, one need only recall the cybernetic view of computers as but one instance of a more general class of automata—automata that had been designed not merely to implement human intentions, not simply as tools for extending our own computational powers, but as machines that could steer, govern, and organize themselves. In short, we need only to be reminded of the new kinds of machine that would embody within their own constitution the kind of telos observed in the development of living organisms.

In one sense it is of course obvious that organisms are not literally to be identified with computers, any more than they are actually to be confused with Wiener's self-steering devices. But the mark of a good metaphor—indeed, the best measure we have of its success—is the very uncertainty we are left to feel about the proximity between its referents. And here, in the legacy of each of these figures of speech, and even more in the legacy of their collective deployment, evidence of just such uncertainty abounds. For biologists of Jacob's generation, and even for those who followed, it had become all but impossible to draw a clear line between regulator and regulated, between genetic information and the quantity that Shannon had earlier called information, between organisms and computers, between digital programs and feedback mechanisms, or between the human intentionality behind and the telos actually embodied within the new machines that had by then so vividly captured our imagination.

Yet, far from exposing a lapse in proper scientific methodology, I am arguing that it was precisely these various kinds of uncertainty that lent the concept of a genetic program its remarkable productivity in the molecular biology of the 1960s and 1970s. Not only did it serve to fill the explanatory gap that had been left by the demise of gene action, by the inability of that discourse to account for development that new research had made manifest, but it also proved valuable on a more local level, and undeniably so: it helped to secure a framework for the hypotheses that early generations of molecular biologists needed to guide their day-to-day research. Indeed, the genetic program can be said to have consolidated the entire family of tropes that guided and gave meaning to virtually all of the discoveries that put molecular biology on the map—the finding not only of regulatory circuits, but also of messenger RNA, the genetic code, translation mechanisms, and even the central dogma of that new discipline. Yet here, too, just as we saw in the case of gene action, productivity also had its downside, and some of that will become apparent in the next chapter, where I examine the cybernetic metaphors of feedback and control in somewhat closer detail, tracing the historical evolution of the particular uses to which they have been put in genetics.

CHAPTER FIVE

Taming the Cybernetic Metaphor

In learning to speak, every child accepts a culture constructed on the premises of the traditional interpretation of the universe, rooted in the idiom of the group to which it was born, and every intellectual effort of the educated mind will be made within this frame of reference.

Michael Polanyi, *Personal Knowledge* (1958)

he term feedback (along with such auxiliary terms as control systems, networks, and steering mechanisms) entered biological discourse in the mid-1950s. The primary context in which this concept arose for biologists—only a few short years before the genetic program—was a coordinated effort among geneticists, molecular biologists, and biochemists to understand the mechanisms of biosynthetic regulation; and its introduction is often taken as evidence of the influence on biological research of modern developments in electric circuitry, electronics, and cybernetics. Certainly, the intense focus of wartime and postwar research on the dynamics of complex, self-regulating systems in systems analysis, cybernetics, computer science, and operations research led to an immensely heightened general interest in both new and older mechanisms of automatic regulation.[1] And at least on a terminological level, the introduction of feedback into biology had surely been influenced by developments in the science of communication engineering, perhaps especially by the immensely popular writings of Norbert Wiener. Furthermore, this debt was widely acknowledged.[2]

Nonetheless, the history of the term's actual function in biological discourse clearly shows that both its conceptual roots and the conceptual consequences of its use in the particular fields in which it was employed were considerably more complex than any simple reading of direct disciplinary incursion might suggest. Indeed, I will argue in this chapter that the word feedback itself, along with the various models that were proposed to illustrate the phenomenon for which it was invoked, constituted a resource for which researchers representing a number of quite different biological agendas competed.[3]

Two of the first uses of the term in the biological literature appear in C. H. Waddington's 1954 paper, "The Cell Physiology of Early Development," and in Frederick E. Warburton's 1955 paper, "Feedback in Development and Its Evolutionary Significance." But probably the most widely cited sources are two papers appearing in 1956 on the inhibition of the synthesis of a metabolite (amino acid or nucleotide) by the end product of the metabolic pathway involved in the production of that metabolite (phenomena that had earlier been called end-product inhibition, blocking, or, more simply, biosynthetic regulation). Edward Umbarger's paper, "Evidence for a Negative Feedback Mechanism in the Biosynthesis of Isoleucine," submitted to *Science* in October 1955, was the first of these; "Control of Pyrimidine Biosynthesis in *Escherichia coli* by a Feed-back Mechanism" by Richard Yates and Arthur Pardee, submitted to the *Journal of Biological Chemistry* one month later, was the second.[4] By 1957, "feedback" had acquired extensive usage in the literature of molecular biology and metabolic regulation.

Despite the significant differences in interest among these various authors, one interest was shared by virtually all, and that was a more or less direct concern with the problem of developmental regulation. The question then, as it remains today, was straightforward: How can one explain the coordinated and regulated process of cellular differentiation in view of the apparent sameness of the genetic complement of all cells? The recent discovery of the structure of DNA had provided an elegant answer to the mystery of genetic continuity, but something else seemed clearly to be necessary in order to account for the generation of cellular difference—in short, to resolve the most conspicuous problem of classical embryology. Of course, studies of metabolic regulation in *E. coli* might seem a far cry from embryology, but much of the interest in the former subject derived precisely from the hope that, in spite of the evident fact that bacterial populations do not undergo cellular differentiation, enzymatic adaptation in bacteria might nevertheless serve as a model for differentiation in higher organisms. As Melvin Cohn wrote in 1958, "Bacterial populations show changes in properties which superficially, at least, appear similar to the phenomenon of differentiation in animal cells."[5] Yet even for those who focused on the comparatively simple phenomenon of enzymatic regulation in bacteria, one can see clear signs of the same tension that prevailed among biologists of the time who thought about differentiation more generally, reflecting different expectations of the directions from which a solution to the problem of differentiation was likely to come.

We may crudely locate the source of this tension in the difference between two leanings, one toward genetic control and the other toward cellular regulation. David Nanney's description of these two conceptual bents, offered at a 1956 meeting on "The Chemical Basis of Heredity," has become something of a classic among historians of biology, and I quote it here:

> The first of these we will designate as the "Master Molecule" concept. This concept presupposes a special type of material, distinct from the rest of the protoplasm, which directs the activities of the cell and functions as a reservoir of information. In its simplest form the concept places the "master molecules" in the chromosomes and attributes the characteristics of an organism to their specific construction; all other cellular constituents are considered relatively inconsequential except as obedient servants of the masters. This is in

> essence the Theory of the Gene, interpreted to suggest a totalitarian government . . . The second concept . . . we will designate as the "Steady State" concept. By the term "Steady State" we envision a dynamic self-perpetuating organization of a variety of molecular species which owes its specific properties not to the characteristics of any one kind of molecule, but to the functional interrelationships of these molecular species. Such a concept contains the notion of checks and balances in a system of biochemical reactions. In contrast to the totalitarian government by "master molecules," the "steady state" government is a more democratic organization, composed of interacting cellular fractions operating in self-perpetuating patterns.[6]

Nanney's advocacy of a reconceptualization in terms of a steady state was not (or at least not on this occasion) intended, as has sometimes been assumed, to argue for the existence of cytoplasmic genes but rather to argue for a more dynamic account of nuclear–cytoplasmic interactions. In fact, he criticized such notions as plasmagenes and cytogenes as a taxonomic "extension of the nucleus into the cytoplasm" (p. 136). Instead, he sought to reverse this trend by restoring heredity to its earlier and more general meaning, redefining the word mutation to refer to any "hereditary modification," and the word cytoplasm to refer to any non-chromosomal (or non-DNA) component of the cell, wherever it might be found (pp. 134–135).[7] "This deliberate confusion of the boundaries between the nucleus and the cytoplasm," he wrote, "is based on a dissatisfaction with a classification of genetic systems in terms of geography alone and on the belief that such systems are more profitably discussed with reference to mechanisms" (p. 136).

The word feedback, at least in its early uses, was well suited to Nanney's agenda—a catch-all term referring indiscriminately to regulatory processes at all levels of the cell or organism, though with a primary emphasis on those processes (of particular interest to Nanney) occurring at the cellular level. However, it soon proved to be a term with such malleability that it could serve in arguments for both kinds of processes, both for genetic mechanisms of control and for dynamical systems of "interacting cellular fractions operating in self-perpetuating patterns." But unlike the example of ambi-valence in the concept of the gene (discussed in Chapter 4), here ambi-valence proved unstable, and in the space of little more than a decade, the multiplicity of uses to which geneticists could put this term seemed to evaporate. By the mid- to late 1960s virtually the only context in which the word would be invoked (at least in the literature of molecular genetics) was in relation to genetically controlled regulation; indeed, feedback seemed to have become an effective synonym for such regulation.[8] How did it happen that its earlier connotations disappeared from view?

Differentiation and Metabolic Feedback

In its first appearance in the biological literature, the term feedback was invoked to describe the "checks and balances of biochemical reactions" responsible for cellular regulation. Waddington (1954) employed it as a descriptor of coupled systems of "autocatalytic processes" or of open systems of chemical reactions admitting of multiple steady states, both of which processes, he believed, could account for differentiation and both of which would lend support to his argument for canalization of developmental pathways.[9] ("Canalization" was the term Waddington employed to refer to the process by which developmental reactions "are adjusted so as to bring about one end result regardless of minor variations in conditions during the course of the reaction.")[10] For Umbarger (1956), the term feedback served a different need: he used it to describe his own very concrete experimental findings in *E. coli*. Manifestly leaning on Norbert Wiener's writings, Umbarger, like Wiener, reached back in time both to Claude Bernard's nineteenth-century concept of the constancy of the *milieu intérieur* and to Walter Cannon's notion of homeostasis from the late 1920s. He wrote:

> In the internally regulated machine, as in the living organism, processes are controlled by one or more feedback loops that prevent any one phase of the process from being carried to a catastrophic extreme. The consequence of such feedback control can be observed at all levels of organization in a living animal—for example, proliferation of cells to form a definite structure, the maintenance of muscle tone, and such homeostatic mechanisms as temperature regulation and the maintenance of a relatively constant blood sugar level. Because of the complexity of so many biological systems, it is often difficult to postulate a mechanism on the molecular level that would serve in a regulatory function.
>
> Less complex systems for study of internal regulation can be found in the orderly synthesis of protoplasmic components during the growth of bacteria.

Writing in the same year (1956), Richard Yates and Arthur Pardee echoed similar sentiments, referring to "the mechanisms by which a cell can control its biosynthetic processes," "cellular economy," and the "control systems" by which bacteria can coordinate their biochemical processes. In none of these early citations was feedback invoked as a mechanism for genetic control.

By the early 1960s, however, the primary locus of feedback, circuits, and control systems in the biological literature had shifted to the specific problem of genetic regulation, and here the fate of the term feedback becomes inextricably intertwined with the discussion in Chapter 4. I return therefore to Monod and Jacob's concluding remarks at the Cold Spring Harbor Symposium on "Cellular Regulatory Mechanisms" in 1961, where they reviewed the implications of the work they had presented earlier in the same proceedings. Their analysis of the regulation of gene activity in terms of the dramatically successful model they had formulated to describe the regulation of β-galactosidase production in *E. coli* (their operon model) had clearly made Monod and Jacob the stars of this symposium, and in their summary of the proceedings they built upon this work to address the more general topic of biosynthetic and cellular regulatory mechanisms—that is, the subject of the conference title.

However, an important difference between regulation in the β-galactosidase system and the kinds of biosynthetic regulation that Umbarger (for example) had analyzed is immediately apparent. In the former case, regulation is effected by an external inducer (galactoside) and an internal repressor (mRNA) and not, as in the latter case, by an end product of the metabolic pathway. In his introduction to the symposium, Davis marked this distinction by referring to Umbarger's system as an example of "feedback inhibition," and Jacob and Monod's system as an instance of "feedback repression."[11] But Monod and Jacob, in a move that was to have great significance, marked the same distinction not by a variation in the noun but by introducing a new term for the modifier of that noun. They relabeled end-product inhibition as "allosteric" inhibition, effectively replacing Umbarger's earlier term of feedback altogether (p. 390).[12] "Allostery" (from the Greek word for "another shape") certainly lacks both the popular appeal and imagistic resonance of feedback, but to biochemists at least, it has clear associations of its own: "steric" refers to the spatial configuration responsible for the activity of an enzyme, and its conjunction with "allo" points to the role of molecules other than those directly affected by enzymatic activity.

The explicit reason given for their introduction of a new term was not that either "feedback" or "end product" inhibition is an inaccurate description of the phenomenon in question but rather that, in these cases, and in contrast to the regulation of β-galactosidase synthesis, the inhibitor "is not a steric analogue of the substrate" with which the synthetic pathway begins. Accordingly, the inhibitor must act not on any of its precursors but on an enzyme required for the conversion of a precursor. "From the point of view of mechanisms," they write, this is what qualifies as "the most remarkable feature" of such effects, and they "propose therefore to designate this mechanism as 'allosteric inhibition.'" Furthermore, they continue, since "there is no obligatory correlation between specific substrates and inhibitors of allosteric enzymes, the effect need not be restricted to 'end product' inhibition. (This in fact is the main reason for avoiding the term 'end product inhibition' in a general discussion of this mechanism)."[13]

But a few pages later in their discussion, a yet more important reason for their reclassification of end-product inhibition emerges. Where their own work described the regulation of protein synthesis, this work involved only the regulation of metabolite synthesis; and from their perspective, it was the former rather than the latter that was of paramount importance. What had been called feedback or end-product inhibition clearly affected the activity of crucial enzymes, but it did not impinge on the synthesis of these enzymes. Thus, their renaming also served to separate the activity of enzymes from their synthesis. From the new taxonomy for biosynthetic regulation that these distinctions carve out, they conclude,

> The models involving only metabolic steady-states maintained by allosteric effects are insufficient to account for differentiation, which must involve directed alterations in the capacity of individual cells to *synthesize* specific proteins . . . The realization that induction and repression are governed by specialized regulatory genes . . . allows . . . the construction of models capable, in principle, of accounting for virtually any type of differentiation. The fact [is] that these mechanisms are not only genetically controlled, but operate directly at the genetic level, and may in some cases quite independent of any metabolic event in the cell. (pp. 399–400)

By their reclassification, "allosteric effects are insufficient to account for differentiation" for the simple reason that allostery, as

Monod and Jacob defined the new term, referred solely to regulation of enzymatic activity, in sharp contradistinction to their own and others' observations of feedback control of enzyme synthesis.[14] Umbarger, by contrast, had originally invoked feedback as an inclusive term—as a description of the inhibition of all biosynthetic processes, and hence encompassing both synthesis and activity. But where Umbarger saw no need to separate either metabolite synthesis from protein synthesis, or synthesis from activity, for Monod and Jacob such a disjunction was crucial: it served precisely to mark the distinction between genetic and cellular regulation.

In their own work on the β-galactosidase system, they had provided a powerful demonstration of regulation of enzymatic synthesis operating at the level of the gene. Furthermore, and thanks to the semantic ambiguity hidden in the very expression genetic regulation (discussed in Chapter 4), their analysis offered an equally powerful invitation to locate the control mechanism for such regulation at the level of the gene (at least as they construed their analysis of that process, and as so clearly indicated in the excerpt quoted above). To be sure, regulation at the cellular level still remained a viable option for other kinds of processes; but by their reasoning, it could not qualify as an explanation for differentiation.[15] They argued that cellular regulation was of a different order altogether: it operated not only in a different realm (the cytoplasm rather than the nucleus) but also by the regulation of different effects (enzyme activity rather than the synthesis of specific proteins). It therefore required both a different mechanism *and* a different name—a mechanism that would allow for the more rapid and more readily reversible kinds of regulation seen at the level of metabolic processes, that is, at the level of enzymatic activity, and a name that would distinguish such regulation from the more stable kinds of regulation operating at the genetic level. Hence their introduction of a new term.

Two different levels of regulation, only one of which could account for cellular differentiation, and the operon model provided a clear mechanism by which genetic regulation could effect a specific change in the state of a cell. Nevertheless, there might well be instances of pseudo-differentiation operating on the metabolic level that could mimic such changes of state. In fact, a number of examples of changes in cell state that seemed to involve no corresponding change in chromosomal genes had already been described in ciliates, and a few geneticists had used these to argue for the existence of cytoplasmic genes. But in Monod and Jacob's classification, there was no room for cytoplasmic genes.

Hence, the question arose, could one account for such instances of "pseudo-differentiation" through allosteric inhibition, operating solely at the level of metabolic processes? Indeed, one could, and twelve years earlier, Delbrück had introduced geneticists to a mathematical model of mutually inhibiting chemical reactions that showed just how it could be done. And now, citing Delbrück's proposal (made "long before feedback inhibition was discovered"), Monod and Jacob invoked that model to underscore the distinctions they themselves wished to draw. Assume two independent metabolic pathways, they explained, "in which the enzymes catalyzing the first reaction in each pathway are inhibited by the final product of the *other* pathway. By such 'crossfeedback' a system of alternative stable states is created where one of the two pathways, provided it once had a head-start or a temporary metabolic advantage, will permanently inhibit the other. Switching of one pathway to the other could be accomplished by a variety of methods, for instance by inhibiting temporarily any one of the enzymes of the active pathway."[16] Yet, for all its superficial resemblance, such a system of alternative stable states must not, in their view, be confused with true cellular differentiation, for it affects only levels of enzymatic activity and not of synthesis.

Monod and Jacob's invocation of Delbrück's model in this context is of interest for at least two reasons. First is the fact that, despite the history of antipathy for mathematical models of development in the American biological community, Monod and Jacob could take for granted that their audience would already be familiar with Delbrück's model.[17] Indeed, it was widely cited in the literature of most immediate relevance—explicitly by Waddington (1954), Nanney (1957), Ephrussi (1956), and Cohn (1956; 1958), as well as in many other discussions of the issues in question—and in fact had been invoked in support of virtually all of the arguments already quoted. But a second point of interest lies in the range of uses to which it had been put. Waddington and Nanney had called upon Delbrück's model to bolster their own arguments for steady-state cellular mechanisms of gene regulation, and within a year, Monod and Jacob would find yet a third use for it: in contrast to its invocation as an example of allosteric regulation, they would employ it to as an illustration of genetic mechanisms for gene regulation. In other words, Delbrück's model was itself deployed in the service of ambiguity; it had become another chip to be used in debates over the locus of regulatory control.

The fact that what is at issue here is not simply a word but a mathematical model provides us with an occasion for asking, What is a model (or theory) a model (or theory) *of,* and what is it *for?* Moreover, that same fact also lends to this example a more general theoretical interest, especially for those interested in the relation between models and metaphors. In the relatively simple view originally advocated by Max Black, metaphors can be seen as a class of models (analog models), both serving the same heuristic value of redescription. Here, however, because it is a model that has become the locus of ambiguity, we can see an instance of models functioning *qua* metaphor, and the function of the model that is highlighted is not so much heuristic as narrative. In the remainder of the chapter, therefore, I want to trace the uses to which this model was put by the various authors who cited it, even reexamining the very literature I have already discussed. Doing so makes it possible to see how its meaning—just what it was intended to illustrate—varies with the immediate purposes of the author, and thereby to observe the changing role it played in the emerging consensus on genetic control.

Genes, Plasmagenes, and Steady States

Delbrück's model was originally introduced in a brief comment at a genetics conference held in Paris in 1949. Compelling evidence had been presented at this meeting for the inheritance of certain phenotypic traits in paramecia and other ciliates through a significant but still limited number of generations. The fact that these traits do not persist indefinitely argued strongly against a chromosomal basis, and G. H. Beale made the case for the obvious alternative, namely, for populations of cytoplasmic genes (or plasmagenes) that are responsible for these traits and that would be transmitted through a finite number of rounds of cell division before being washed out. Delbrück's comment was offered in direct response to Beale; his aim was to show that it was possible to account for these observations in yet a third way, as a purely epigenetic phenomenon requiring no invocation of either nuclear genes or plasmagenes.

In a note entitled "Enzyme Systems with Alternative Steady States," Delbrück wrote: "The point I wish to explain is this: many steady state systems are capable of several alternative steady states under the same environmental conditions. They may switch from one to another steady state under the influence of transient perturbations."[18] To illustrate such a possibility, Delbrück proposed a particular model of cross-reacting metabolic pathways (see Figure 4). Without offering any of the mathemati-

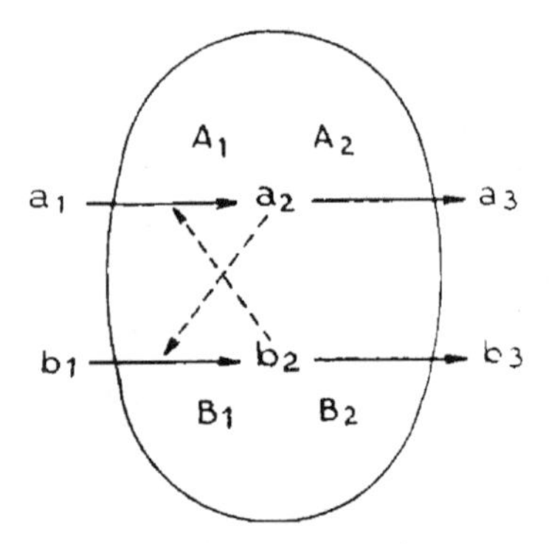

Figure 4. A_1, A_2, B_1, B_2 represent different types of enzymes within the cell; a_1 and b_1 represent foodstuffs in the medium, acted upon by enzymes A_1 and B_1 to produce intermediate metabolites a_2 and b_2, which are in turn acted upon by A_2 and B_2 to produce waste-products a_3 and b_3. In a constant environment the cell quickly reaches a steady state. (Delbrück, 1949b.)

cal detail, he argued that the dynamics of this system permit the cell to exist, in the same environment, "in two extremely different functional steady states, without any change in nuclear genes, plasmagenes, enzymes or any other structural biological units, and transitions from one state to the other can be brought about by *transient* environmental changes" (p. 4). Furthermore, as he also noted, "The theorem which I have indicated is not new" (p. 4).

Indeed, for those with mathematical competence, the possibility of multiple steady states for interacting chemical reactions should have been a relatively straightforward consequence of the nonlinear coupled equations for such reactions, and it had already been pointed out a number of times.[19] But it was an entirely novel idea to geneticists,[20] and even without the supporting mathematics, it was quickly seized upon by critics of the plasmagene hypothesis with the hope, as Horowitz and Mitchell put it, "that development may eventually be accounted for without the assumption of plasmagenes"—that is, that nuclear genes would suffice.[21] At the same time, however, it was also deployed by erst-while defenders of extra-nucleic inheritance who used it to clarify and extend their own earlier arguments for non-chromosomal (or cytoplasmic) agency. Nanney, for example, invoked Delbrück's model (as well as Wright's more general description of the same kind of phenomenon) as an example of "self-perpetuating metabolic patterns," illustrating how "antagonistic biochemical pathways provide a mechanism both for integrating cellular activities and for transmitting specificities in vegetative growth."[22] In short, even from the beginning, a certain plasticity can be identified in the programmatic uses to which this model lent itself.

In the mid-1950s, with the introduction of the concept of feedback into the literature, Delbrück's model found yet more extended usage. Waddington, more mathematically sophisticated than most biologists, was one of the first to see the connection with feedback, and hence to see a still different kind of argument for which Delbrück's model could be useful, one that would not be in conflict with Nanney's use but would be more directly geared to his own advocacy for canalization in developmental pathways. In the same article in which he introduced the term feedback to support his argument for canalization, Waddington invokes Delbrück's model of direct interaction between synthetic pathways as an example of the kind of feedback mechanism his view of differentiation required.[23] He also uses it to enter a plea for the "theoretical outlook which we require to understand the mechanisms of differentiation. The field [of developmental biology] would, I think, repay much more study than it has yet received . . . I hope that some competent mathematician, at home in the field of chemical kinetics, will interest himself in it."[24]

Waddington's plea for more mathematical study may have gone unheeded by geneticists, but Delbrück's verbal description of the insights such study might offer did not. It made its debut in discussions of biosynthetic regulation in *E. coli* at an international symposium held in November 1955 entitled "Enzymes: Units of Biological Structure and Function."[25] Here, Delbrück's model is cited both by Boris Ephrussi (p. 37) and by Mel Cohn (p. 46). Ephrussi, after skeptically observing that, despite the ingenuity of the models proposed, "concrete examples of alternative steady states of structurally identical biological systems, initiated by differences in environment but perpetuated in identical environments, are probably even rarer than cases of cytoplasmic heredity," immediately bows to Cohn for just such a concrete example. Following Ephrussi's talk, Cohn then introduces Delbrück's model to support his explanation of the inhibition of induced synthesis of β-galactosidase. He writes: "The glucose-inhibition effect is the first example of a well-characterized system which might fit the Delbrück model, and it might be fruitful to see eventually how many other cases of memory, e.g., long-term adaptation, may be described in terms of systems of the y-β-galactosidase type which are inherently autocatalytic and self-stabilizing" (p. 46). Two years later, the biochemistry of this system has been more clearly elucidated and Cohn is more confident: "The β-galactosidase system is then the first analyzed example of the type of steady-state model proposed by Delbrück some ten years ago to explain the appearance of a stable 'new' character in cells without invoking genotypic changes."[26]

By 1961 Monod and Jacob had developed their operon model for the β-galactosidase system, which they saw as clearly indicating regulation at the level of the gene. They too invoke Delbrück's model, but not for β-galactosidase repression.[27] Consistent with their recasting of feedback inhibition as allosteric inhibition, a mechanism referring solely to inhibition of enzyme activity, and hence referring to regulation as operating only on the cellular level, Delbrück's model—being a model for *cellular* regulation—is here recruited to illustrate the properties of allostery (and incidentally claimed as a historical precursor to the discovery of feedback inhibition).[28] Their recourse to Delbrück is in this sense consistent with his own original use of that model, but the added implication that "models involving only metabolic steady-states maintained by allosteric effects are insufficient to account for differentiation" is their own. One year later, however, their argument (and their use of this model) undergoes a curious shift.

In June 1962, addressing the Asilomar meeting of the Society of Developmental Biology, Jacob and Monod focus their attention directly on the mechanism of differentiation in higher organisms. Their audience on this occasion consists largely of developmental biologists, many of whom have not yet been won over to the strengths of the new molecular biology. As might therefore be expected, they open their presentation with a general description of regulatory phenomena with which few could object: "The complex and precise chemical network of information transfer upon which the development and physiological functioning of organisms must rest, implies the existence of precise regulatory systems at the level of both the organism and the cell."[29] Very quickly, however, they move on to their particular example of genetic regulation, and their question becomes, "Can the basic elements of regulatory circuits found in bacteria, i.e. regulator genes, repressors, operators, be organized into other types of circuits, whose properties could account for the main features of differentiation in higher organisms?" This question, they continue,

> deals with the mechanisms which may account for the orderly emergence of differentiated functions. The stability or clonal character of differentiation point to "hereditary" phenomena, but the main problem . . . is concerned with the nature of these phenomena . . . This problem must be defined in chemical terms, namely, whether differentiation involves conservation of, or specific changes in, the information coded in the base sequence of polynu-

> cleotides. Change implies any mutational alteration in the sequences, or any distribution of sequences contained in chromosomes or in extranuclear elements. Conservation implies differential functional activity of nucleotide sequences, resulting, for instance, from the establishment of steady state systems capable of clonal perpetuation, as pointed out by Delbrück (1949). (p. 52)

Only a year earlier, Monod and Jacob had argued that "models involving only metabolic steady-states maintained by allosteric effects are insufficient to account for differentiation," that only genetically controlled regulation—operating directly at the level of the gene—can do the job. But now, just such a model is invoked as a mechanism for the "clonal character of differentiation"; furthermore, it is employed not (as it had earlier been) as an illustration of the regulation of enzymatic activity but of regulation of enzyme synthesis at the level of the genes, that is, of regulation of the "differential functional activity of nucleotide sequences."

This shift in Monod and Jacob's use of Delbrück's model occurs in the context of a new framing of the division between the terms genetic and epigenetic in ways that can no longer be made to neatly align either with a division between genetic and cellular or with their earlier dichotomy between regulation of enzyme activity and of enzyme synthesis. In place of the earlier dichotomy between activity and synthesis, we now have a distinction between informational change and conservation, and because of this reframing, both the conceptual and topological boundaries need to be redrawn. The very terms of the new configuration reflect a shift from the discourse of biochemistry (and also, though less conspicuously, from the taxonomy of classical genetics) to the language of molecular biology, and specifically, to the language of Francis Crick's central dogma. In 1958 Crick had proclaimed, "Once 'information' has passed into protein, it cannot get out again."[30] In fact, genetic information must not be modified in the course of development, for its conservation in the DNA is what guarantees the stability of intergenerational heredity. In these new terms, the question of differentiation is no longer one of a debate over genetic versus cellular control but rather over the question "Does differentiation in higher organisms require *any* change (or redistribution) of the information coded in the DNA?" Or, as molecular biologists clearly believed to be more likely, "Can it be accounted for within a purely conservational scheme (that is, within the central dogma), invoking only 'differential functional activity of nucleotide sequences'"? Furthermore, with the submergence of earlier (topological) distinctions between genetic and cellular control, or between regulation of enzymatic activity and synthesis, and the reframing of the question in terms of conservation, Delbrück's model takes on a different value. Precisely because it fits within a purely conservational scheme, it can now qualify as a mechanism for differentiation, and Jacob and Monod can invoke it as an instance of a new category of epigenetic (yet genetically controlled) mechanisms for regulating the synthesis of the specific proteins required for differentiation. In other words, it now belongs to the same genre as their own operon model—a genre that no longer opposes cellular to genetic regulation but rather assimilates the former to the latter.

The recategorization of Delbrück's model is in fact important to the new logic of their argument, for that model provides just the plausible mechanism they now claim has been lacking for cellular regulation, but which is available for genetic regulation. As they continue,

> Although frequently favored in past years, [the hypothesis of systematic modifications of the information contained in the genetic structure] still lacks the support of experimental evidence as well as a plausible mechanism to account for an orderly distribution of cytoplasmic particles.
>
> Opposition to the alternative models, i.e. stable activation or inactivation of chromosomal segments, has come mostly from the difficulty of visualizing a suitable mechanism able to alter the func-

> tion of genes without altering their informational content. The study of genetic regulation in bacteria provides just such a system.

And by way of a conclusion, the authors write: "All these facts encourage the hypothesis that differentiation operates at the genetic level, using elements basically similar to those found in bacteria." If true, "if differentiation is based on genetically controlled circuits, then genetic analysis of somatic cells may well turn out to be essentially an analysis of gene expression, as controlled by gene interaction."[31]

Many years have passed since 1963, and, not surprisingly, the frequency with which the Delbrück model is cited in the literature on biological regulation has declined. To the extent that it endures, that model is far more likely to be recalled for the final use to which Jacob and Monod put it than for the initial uses for which it was invoked during the 1950s, that is, it is used as a model for genetic regulation rather than as a model for cellular regulation. The person who has contributed most to keeping it alive is almost surely René Thomas, a molecular biologist who had worked with Monod and Jacob in the early sixties and has since that time devoted much of his energies to analyzing the logical structure of genetic control circuits. In a recent review of this work, Thomas and D'Ari credit Delbrück's "epoch-making remark" as the first suggestion of epigenetic regulation.[32] Notably, however, the discussion that follows focuses primarily on regulation at the level of the gene. Although the models that have subsequently been developed by Thomas and others are in fact quite general and eclectic in their application, only in passing do the authors note the limitations of a focus on transcriptional regulation: "If regulation is exerted at the level of mRNA modification, splicing, translation, or posttranslational events, these features can be included in a more detailed description with no particular difficulties" (p. 5). As to the persisting problem of identifying the locus of control in regulation, at whatever level that regulation is seen to operate, no mention is made at all.

Models and Metaphors

The ten years that followed Watson and Crick's spectacular announcement in 1953 stand unrivaled in the history of twentieth-century biology for the sheer drama of the discoveries they witnessed. By the end of that period, a clear picture had been formed of the chemical structure of the genetic material and the mechanism of its replication, of the process by which the sequence of nucleotides is transcribed and translated into sequences of amino acids, and of at least one mechanism by which the rates of synthesis of different proteins are regulated. We also had gathered vastly more information about the biochemical intricacies of the metabolic pathways that keep the machinery of life going. But the research on which such clarity depended was of necessity a process of groping—groping for words, for concepts, for new experimental handles. It could not have been otherwise.

That a word can support a number of different constructions is by now a familiar story, but the same kind of versatility in the uses of a mathematical model is rather less so. A mathematical model may of course be applied in any number of different contexts, but at least in physics and engineering its primary function is conventionally taken to be the enabling of predictions and the guiding of experimental research in the immediate context in which it is employed, whatever that context may be. Take, for example, the familiar model of the harmonic oscillator as represented by the equation $d^2x/dx^2 = -kx$. This equation may be used to describe the behavior of a pendulum, a spring, a hydrogen atom, a string in tension, a one-dimensional plasma, or any number of other phenomena. In all these uses, its value is measured by its adequacy as a description of experimentally controlled observations of the particular phenomenon in question.

This, however, appears not to have been the case with the Delbrück model; the uses to which it was put in biology were conspicuously different. Despite (or perhaps even because of) its popularity, its role seems to have been less one of suggesting new research and more one of bolstering existing research programs—programs that not only were well articulated but continued without any evidence of being substantively influenced by any quantitative or qualitative predictions that model might make. Even its value of establishing plausibility for particular kinds of processes was mitigated by the variety of such processes for which it could be invoked. One might thus say that its primary role was rhetorical. And in this role, it revealed much the same kind of plasticity as did the term feedback—in fact, just the kind of versatility we have come to expect from any figure of speech.

Of all the variability in the uses of Delbrück's model—as counter to the hypothesis of cytoplasmic genes; as support for an expanded notion of cytoplasmic heredity (and challenge to the master-molecule orientation of molecular biology); as support for canalization; or, more generally, as an example of cellular regulation—by far the most significant shift is that which shows up in Jacob and Monod's 1962 reinterpretation. Here, the very same model is cited as a regulatory mechanism conserving the informational content of the DNA, and hence supporting the central dogma of molecular biology. In its new role, Delbrück's model becomes an instance not of cellular regulation but of genetic regulation, exemplified par excellence by the operon.[33]

Certainly, the operon model was a tour de force, one that contributed decisively to shifting the dominant discursive and conceptual framework of genetics from the study of gene *action* to that of gene *activation.* And because it offered a possible resolution of the paradox of dividing embryology from genetics, it was immediately embraced by many embryologists. The noted embryologist John Moore was one of these, and in his 1963 overview, *Heredity and Development,* he wrote: "This [operon] hypothesis is consistent with the thinking of embryologists who fail to see how a genetic system identical in all cells, alone provides for cellular differentiation . . . Though the genetic system specifies what a cell may do, non-genetic phenomena influence what it actually does. This point of view, which once would have been reasonable to an embryologist but not to a geneticist, now seems reasonable to both." And a few pages later, he adds, "A generation ago few embryologists or geneticists would have predicted that a synthesis of their fields would be made possible by studies on the bacterium E. coli. But this microscopic creature, with no embryology of its own, has shown a way. A decade from now it may be difficult to distinguish between a geneticist and an embryologist."[34] But notice: Moore's reading of the actual model, as well as of the promise for rapprochement it offers, is quite different from that of Jacob and Monod's. For Moore, the operon model provides an instance not of genetic control but of developmental control. Appearing in a chapter entitled "Developmental Control of Genetic Systems," and in keeping with that title, his remarks suggest a sharp distinction between the genetic system that merely "specifies what a cell may do" and the "non-genetic phenomena [that] influence what it actually does."

But whichever way claims for the synthesizing power of the operon model were read, over the long run all such claims—made by embryologists or by geneticists—would prove to be premature. Monod had been too confident. What was true of *E. coli* turned out, after all, not to be true of the elephant.[35] For that long-awaited synthesis between the two fields, for the day when it would become "difficult to distinguish between a geneticist and an embryologist," biology would need to wait still longer—at the very least, until molecular geneticists had begun to turn their experimental attention to more complex organisms than *E. coli* (and especially, to those organisms that actually undergo the systematic changes required by ontogenetic development).

But let us return to the 1960s and to the topic at hand. Because of their work on the operon model, Jacob and Monod have also been credited with introducing feedback, regulatory circuits, and epigenetic processes into molecular biology, and for good reason. It should be noted, however, that even while popularizing these concepts among their colleagues, they also tailored them in ways that minimized conflict with the dominant framework of their time. Where others had taken the essence of the cybernetic vision to lie in the absence of any form of centralized control, and had seen feedback as providing the mechanism by which global regulation could be achieved, Jacob and Monod redefined both feedback and regulation (and even epigenesis) to refer to genetically controlled processes. The possibility of such redefinition was, of course, a direct consequence of the instability already inhering in all these terms, and redefinition did help to stabilize their meaning for geneticists. But more importantly, in taming the cybernetic metaphor, it also provided a way to put that metaphor to use even while bracketing (and even obscuring) the challenge an alternative reading would have offered.

To be sure, these were early days in the study of regulatory mechanisms, and Monod and Jacob could hardly be expected to have anticipated the variety of mechanisms that have subsequently been found, operating not only on the level of genes but also on the level of chromatin structures, the RNA transcript, the translation process, and even protein structure. Especially, they could not have anticipated the complexities of the regulatory mechanisms that would soon be found in higher (eukaryotic) organisms. Far from conserving genetic information, some of these mechanisms can actively alter the informational content of the genes, and do so in ways that are themselves closely regulated, responding to the changing needs of the developing organism. (I refer, for example, to such processes as alternative splicing, post-transcriptional editing of mRNA, or enzyme-mediated modification of protein composition.)[36] Still, their reading of the operon model as evidence not only of regulation at the level of the gene but of regulation controlled by genes, and their insistence that only such mechanisms could account for differentiation—however plausible that reading may have been to geneticists—did have its downside. Given its rapid acceptance by the vast majority of readers, it inevitably had the effect of discouraging further exploration of extra-genetic (or epigenetic) mechanisms of regulation. For the next two decades, cellular regulation would in the first instance *mean* genetic regulation to most molecular geneticists, and what little research had begun on such extra-genetic mechanisms of regulation as metabolic networks would come to a virtual standstill.

Yet even so, it would be a mistake to assign Monod and Jacob too much responsibility. Certainly, what they argued mattered, especially given the success of their model and the magnitude of their influence on molecular (and even on developmental) biology in the 1960s and 1970s. But both their own reading of their work and its prompt acceptance by their colleagues were themselves a product of (or preconditioned by) the assumptions and expectations of the mindset which they and their readers had inherited. And the primary vehicle by which mindset is transmitted is of course language. Viewed in these terms, their reading might be seen as indicating a principle of parsimony in the evolution of scientific thought—a prescription for conserving as much of the pre-existing formulations as new findings would permit. The successes of the first two decades of molecular biology were staggering, and to the extent that they vindicated the beliefs and expectations of classical geneticists, they also endowed those same beliefs and expectations with new strength. Thus, we are drawn back to the most familiar of all metaphors in the history of genetics, for language itself operates here as a conservative force. Much as with genetic structures, language builds into new formulations a tacit memory of older concepts, shaping the course of research in accordance with its prior history, even while it also, and at the same time, provides the means by which new concepts are formulated and new perceptions achieved. Words—together with the linguistic forms by which they are given meaning—are, in this sense, just like genes and the regulatory networks in which they are employed, the primary vehicles for their own evolution.

CHAPTER SIX

Positioning Positional Information

Mihi a docto Doctore
Domandatur causam et rationem quare
Opium facit dormire:
À quoi respondeo,
Quia est in eo
Virtus dormitiva,
Cujus est natura
Sensus assoupire.*

Molière, *Le Malade imaginaire* (1673)

olecular biologists began to show visible signs of restlessness in the late 1960s. The golden days were over, or so it seemed to many observers. In 1968 Gunther Stent published an article called "That Was the Molecular Biology That Was," in which he wrote of "the approaching decline of molecular biology, only yesterday an avant-garde but today definitely a workaday field."[1] Stent was not alone. A number of the original pioneers set out in search of new frontiers where, buoyed by their successes with *E. coli,* they took on the challenges of working out a genetics of behavior, of neurophysiology, or of development in higher organisms. That very year, Seymour Benzer turned his attention to *Drosophila,* Sydney Brenner to *C. elegans,* George Streisinger began to explore the uses of Zebra fish as a new model organism, and a number of others (for example, David Hogness, John Gerhart, and Mark Kirschner) took advantage of sabbatical leaves to visit laboratories where

* "How does opium induce sleep? By virtue of a faculty, the nature of which is to tranquilize the senses."

they could learn about the traditional model organisms of embryology and acquire training in more classical methods.[2] Exposure to the complexities of development in higher organisms proved eye-opening. Out of the new alliances that were formed, and out of the cross-fertilization between molecular and classical techniques of analysis that resulted, came the rebirth of developmental biology as an active field for research.[3]

Just a few years later, starting in the mid-1970s, the infusion of new techniques of recombinant DNA (making it possible to target, disrupt, recombine, and clone individual genetic elements) opened up a range of investigative opportunities heretofore undreamed. Indeed, many observers regard the introduction of such techniques as the most critical factor precipitating the emergence of a new molecular developmental biology.[4] It is difficult to untangle the relative importance of events that occurred so closely in time and that were so quickly assimilated with one another, but what is not in dispute is the basic fact that over the last quarter of a century we have witnessed a dramatic convergence of genetics, embryology, and molecular biology, resulting in a radically new understanding of some of the mechanisms involved in ontogenetic development. And along with that new understanding, a new figure of speech has appeared in explanations of development—one that, in much the same ways as the earlier figures I have discussed, drew its power from its inherent flexibility. This was positional information.

When the notion of positional information (PI) was first introduced, it was clearly intended as a companion to (and supplement of) the by this time well-established term genetic information. The concept was formulated in an effort to account for intercellular regulatory phenomena for which intracellular (or genetic) mechanisms were thought not to be immediately relevant, and hence as a direct complement (if not an alternative to) the concept of a genetic program.[5] The specific problem which the genetic program had been designed to address was the regulation of gene expression: How do cells with the same genes develop so differently? Yet differential gene expression was hardly the only challenge that developing organisms posed for geneticists. At least as baffling as the regulation of gene expression in individual cells was the phenomenon of spatial self-regulation in the organism as a whole—how it happens that organisms develop with the right parts in the right place, in the correct proportions, more or less independent of the size of the whole.

In fact, this problem had plagued biologists ever since Hans Driesch's famous demonstration in 1892 of the development of normal, albeit smaller, plutei from isolated blastomeres of a sea urchin embryo, and it was particularly vexing for those who studied regeneration. As C. H. Waddington put it, "One of the most striking characteristics of embryos is that they 'regulate'; that is to say, if pieces are cut out of them or they are injured in various ways, they have a great tendency nevertheless to finish up by producing a normal end result."[6] The parts of a developing embryo—possibly even the individual cells—seem to "know" where they are, and the obvious question is: How do they acquire this knowledge? How could cells know where they are? Surely not by their genes, for genes do not carry spatial information. In fact, ever since the time of Driesch, morphogenetic phenomena had seemed so utterly to defy explanation in terms of discrete hereditary particles that an alternative framework of fields and gradients had emerged.[7] But with the rising influence of a genetic perspective on development in the middle part of the century, both the subject of morphogenesis and the alternative framework with which it had become associated gradually fell into disrepute. John Opitz writes, "In one of the most astounding developments in Western scientific history, the gradient-field, or epimorphic field concept, as embodied in normal ontogeny, and as studied by experimental embryologists, seems to have simply vanished from the intellectual patrimony of Western biologists."[8] Indeed, for most developmental biologists working in the 1960s, even to speak of parts of an organism (or cells) as knowing seemed to invite a vitalistic reading of development; and by this time, the problems posed by such phenomena had largely receded from view. And the very question of the genesis of organismic form was seen as belonging to another domain, if not another age altogether.

By the middle 1960s, virtually the only areas in which interest in morphogenesis, and particularly in phenomena of spatial regulation, seemed to survive were those of regeneration studies and theoretical biology. Given their shared interest (and perhaps also given their shared sense of marginalization), some kind of bridge between these two very different fields seemed natural, and it was in the context of this bridge that Lewis Wolpert first introduced the actual term positional information in the late 1960s. Without question, the term was both a convenient and opportune way of referring to the fact that cells seem to know where they are; for Wolpert, however, it also designated a theory, an account of the means (or mechanism) by which cells come by this knowledge. Indeed, ever since its debut, the term has been plagued by a persistent uncertainty as to whether its referent is a phenomenon or an explanation of that phenomenon. This uncertainty is the subject of the present chapter. With even greater insistence than the terms I have discussed in the previous two chapters, the history of PI provokes the question: What's in a name?

The Origins of Positional Information

At the time this story begins, Wolpert was a relative newcomer to developmental biology. Originally trained as a civil engineer in South Africa, he had completed a Ph.D. on the mechanics of cell division in 1960, followed this with post-doctoral work in

Sweden on gradient models for development, and by the mid-1960s was fully absorbed by the curious patterns of hydra regeneration. C. H. Waddington, seeking to revive the project of a theoretical biology, organized and convened the first of a series of annual symposia on that subject in 1966, and it was here that Wolpert presented his first attempt at "fitting the process of development into a general theoretical framework."[9] This was the "French Flag" model—invented, as he later wrote, "in order to formulate the problem of pattern formation rigorously." Imagine, for example, three possible cell fates (corresponding, say, to the three colors of the French flag), where each fate (or color) is specified by a fixed threshold in concentration (see Figure 5). If we now compare two embryos of different length, we need nothing more than the law of similar triangles to see that the proportions of the embryo are preserved—in other words, that spatial regulation has been achieved.[10]

By itself, however, the French Flag model did not yet include an answer to the question of how cells know where they are. Wolpert's answer—what he and others now think of as his crucial contribution—came with the formulation of the notion of PI. He presented this concept two years later at Waddington's third symposium, where Wolpert delivered the first of a long series of papers entitled "Positional Information and Pattern Formation."[11] "Positional information," he now proposed, "is the main feature which determines the pattern of cellular differentiation, and . . . the mechanism of position determination is universal. To put it bluntly a cell knows where it is, and this information specifies the nature of its differentiation." His hope, expressed in his concluding remarks, was that this concept "will have provided a useful unifying framework and will give new meaning to such concepts as gradient, induction, dominance, and field."[12]

More than twenty years later, Wolpert recalls that "the idea of positional information came to me in early 1968. It was a very

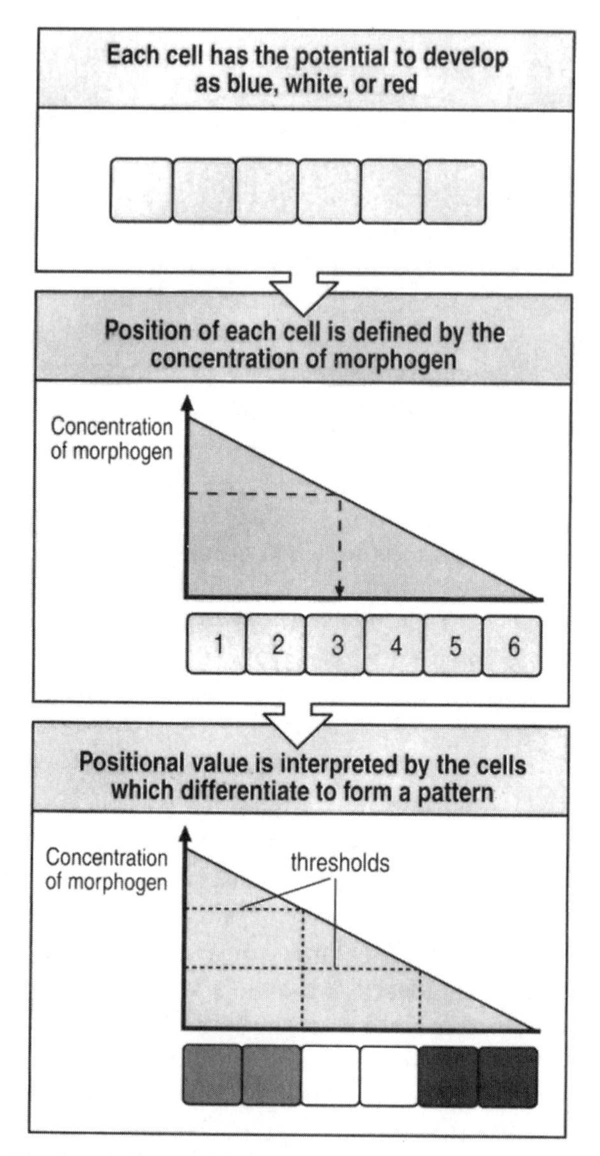

Figure 5. The French Flag model of pattern formation. Each cell acquires a positional value defined by the concentration gradient of some substance at that point, and differentiates into blue, white, or red according to its interpretation of its positional value. The basic requirement of the model is that the concentration at the two end points remains constant. (Wolpert, *Principles of Development*, 1998, © Current Biology, Ltd.; rpt. by permission of Oxford University Press.)

exciting few days when everything became clear and obvious." But he also recalls that not everyone shared his enthusiasm: "In the summer of that year I was at Woods Hole. I presented the new ideas at a Friday evening discourse, which had a very large audience. The reception was very hostile. They did not like being told that for the limb, e.g., they had completely missed the problem . . . Only Sydney Brenner was encouraging."[13] And in a personal interview, he adds, "No one talked to me; they wanted to know 'Who do you think you are?' "[14]

Indeed, who *was* Lewis Wolpert? By the late 1960s his name was familiar to those interested in theoretical or mathematical biology, to members of the hydra community, and even to theoretically inclined students of regeneration.[15] But outside these small enclaves, it would scarcely have been known at all. Furthermore, given both the traditional antagonism to abstract theory among most experimental biologists and the ostensibly extra-genetic thrust of Wolpert's argument, the hostility he met is hardly surprising. Yet, from this inauspicious beginning, Wolpert's subsequent rise to prominence has been spectacular. In a recently published interview, he is described as "one of the most influential developmental biologists in Britain and the world," and his concept of PI is claimed to have "changed the way we think about pattern formation in the embryo and allowed new generations of molecular developmental biologists to frame their questions in a way that would give sensible answers."[16]

Without question, the notion of positional information has grown steadily in popularity over the intervening years, and the term has by now become a staple in the vocabulary of developmental biologists. A very crude estimate of its increasing currency in the biological literature over the last three decades can be obtained by searching Medline. Over the years 1966–1970 it shows up as a keyword exactly once; from 1971–1975, in 33 articles; from 1976–1980, in 44; from 1981–1985, in 82; from 1986–1990, in 139; from 1991–1996, in 273; and from 1996–2000, in 323. To be sure, Wolpert himself has been an exceedingly vigorous promoter of its power to provide a unifying framework for understanding development, and the conference held in London in September 1996 to celebrate its upcoming thirtieth anniversary was organized by former students and co-workers.[17] But the Medline citations clearly show that the popularity of both the notion and the term has come to extend well beyond Wolpert's immediate community. From her review of the literature, Lee Zwanziger concludes that, because of his introduction of this concept, Wolpert is credited "with substantially reformulating the inquiry [embryological pattern formation] into the central problem of developmental biology."[18] And speaking as a representative of the new molecular developmental biology, Elliott Meyerowitz has publicly referred to PI as one of the two theoretical cornerstones of molecular developmental biology (the other being the central dogma).[19] The question, of course, is: What exactly does "positional information" mean? Also, what kind of theoretical cornerstone does it provide? And how does it help us to understand development?

Two features of this story stand out as being of particular interest for a discussion of models and metaphors. First is the incorporation of an abstract concept that was originally intended as a corrective to gene-based accounts and introduced in support of an avowedly theoretical (or mathematical) model of development into the mainstream of experimental genetics. And second is the transformation this concept has undergone in the course of its history. Indeed, I will argue that these two features are closely linked: that it was precisely the transformation of the meaning of PI—in ways that brought that concept into ever closer conformity with developments in molecular genetics—that made its incorporation into molecular genetics possible. When Wolpert first introduced the term, the only visible prospect of a reconciliation between genetics and embryology was provided by Jacob and

Monod's operon model of gene regulation in *E. coli.* Molecular analyses of development in higher organisms had not yet begun, and no one then could have foreseen the new avenues of reconciliation these analyses would open up. The rise to prominence of PI over the intervening three decades has depended critically on the shifts in perspective the new molecular developmental genetics has brought. But at the same time, that rise in prominence has also depended—and equally critically—on the conceptual malleability of the term and on the opportunities this malleability afforded to adapt its meaning in ways that would make it applicable to the new contexts that were emerging.

In its original formulation, PI simply referred to a (or the) mechanism by which cells estimate their location within the body of the organism.[20] The physiological basis of such a mechanism was of course unknown, but given its assumed universality, Wolpert considered a genetic approach to be "not very promising." Since "any change in its specification would drastically affect all systems," he argued, "the possibility of viable mutants seems extremely unlikely."[21] For an alternative to gene-based intracellular mechanisms, he looked to the kind of general abstract framework that had long been associated with speculations about fields, gradients, and spatial regulation. Indeed, Hans Driesch had made an apparently similar suggestion in 1908 when he wrote that "the prospective fate of any blastula cell is a function of its position in the whole."[22] And, at roughly the same time, T. H. Morgan had suggested that the specification of embryonic polarity might be due to the pre-existence of a spatial "chemical or physical" gradient (1905). Nevertheless, in Wolpert's view, PI was far more than a new name for old ideas. Above all, it provided the "specificity" and "rigor" that such earlier hypotheses had lacked. He was particularly intent on distinguishing his own theory from Charles Manning Child's efforts to follow up on Morgan's speculation, and from the ignominy into which those efforts had subsequently fallen: "As Spemann (1938) has pointed out, the gradient theory of Child failed to provide a mechanism whereby quantitative differences were translated into pattern."[23]

Yet it is just here, in the meaning attached to words like specificity, rigor, and mechanism, that we can see the magnitude of the epistemological gap separating Wolpert in the late 1960s from most of his contemporaries in experimental genetics, molecular biology, and developmental biology. What in his view lent positional information specificity and rigor, what made it a mechanism, came in the first instance neither from experimental observations nor from its attachment to any plausible physiological mechanism but rather from its association with the surprisingly simple mathematical model Wolpert had introduced to represent the developing embryo, that is, his French Flag model. In fact, this model was so simple that it could be understood by anyone equipped with high school–level proficiency in mathematics.

In its most rudimentary form, it represents the spatial structure of the embryo by a linear gradient in the concentration of an unspecified substance (a morphogen) in which the concentration at the two ends of the embryo is maintained at constant levels, independent of its length, by hypothetical sources and sinks. A cell "knows where it is" by measuring the concentration of this substance at the point at which it finds itself; in other words, the information specifying its fate is to be found in that number.

The sheer simplicity of the model was at once an asset and a liability. On the one hand, it could appeal to those who, like Susan Bryant, were "non-mathematical, but theoretically minded." But to many others, the notion that a system as complex as a developing embryo could be reduced to so simple (even simplistic) a scheme was not only unthinkable but downright insulting—both to the embryo and to themselves. Wolpert himself recognized the implication of his presentation that "they had completely missed the problem."

Also, there were other difficulties as well. In the absence of any recognizable connection between genetic information and positional information, geneticists attending his Friday night lecture in Woods Hole might have considered him to be speaking a foreign language, and molecular biologists in the audience would surely have been perplexed by his use of the term mechanism. What, they would have wondered, is the actual physical or chemical mechanism by which this sort of positional specification might arise, and where is the evidence for such a mechanism?

Given this initial skepticism, how are we to understand the mounting popularity of PI over the ensuing years? What changed? Two different kinds of changes seem to be crucial, one of which is to be found in the transformation the concept underwent in Wolpert's reach for wider audiences, and the other in the metamorphosis in molecular approaches to the study of development that we have witnessed in recent decades.

Beginning with the first of these, we notice that the specific attachment of PI to the French Flag model has gradually receded from its initial primacy in Wolpert's advocacy of the concept, and so too has its formulation as a complement or alternative to genetic mechanisms. In the early period, he is explicit on this point. For example, in 1971, he writes:

> Does the genetic material provide a description of the adult? What, one may ask, are the genes for leg formation in tetrapods, and how do they make a leg? Or what are the genes for gastrulation? A current fashion in molecular biology is to suggest, either explicitly or implicitly, that the answers to such problems will come from deeper and deeper molecular probings. Characterize the RNA and proteins and the form will look after itself, or at least be immediately explicable. Such a view suggests that if we understood cytodifferentiation or molecular differentiation then pattern would be explicable. I wish to take a rather different view and would suggest that the development of form and matter, while related to molecular differentiation, can be viewed in their own right. Moreover, the rules, laws, or principles for the expression of genetic information in terms of pat-

> tern and form will be as general, universal, elegant and simple as those that now apply to molecular genetics.[24]

By 1975, however, he is noticeably more accommodating. In a paper co-authored by J. H. Lewis, he now writes, "A theory of development would effectively enable one to compute the adult organism from the genetic information in the egg." At this point in time, the only difference between his approach and that of molecular genetics has become one of strategy, of different ways of viewing a single phenomenon, and he continues:

> The problem may be approached by viewing the egg as containing a program for development, and considering the logical nature of the program by treating cells as automata and ignoring the details of molecular mechanisms. It is suggested that development is essentially a simple process, the cells having a limited repertoire of overt activities and interacting with each other by means of simple signals, and that general principles may be discerned. The complexity lies in the specification of the internal state which may be described in terms of a gene-switching network.[25]

With the tacit conflation between "the genetic information" referred to in the first sentence and "a program for development" in the second, earlier implications of a material distinction between a genetic program inscribed in the DNA and a developmental program contained in the egg have been submerged if not altogether erased, as has the suggestion that deeper molecular probings into the structure of the genetic material might not suffice to account for morphogenesis and pattern formation.[26] And perhaps wisely so, for the mid-1970s marked the beginning of the dramatic breakthroughs in molecular genetics that have so altered the character of that endeavor, and that have more specifically permitted such radically new insights into the very problems with which Wolpert was concerned.

Of particular importance was the commencement in 1975 of the collaborative investigations of Christiane Nüsslein-Volhard (originally trained as a molecular biologist) and Eric Wieshaus (trained in embryology) into the genetics of embryonic development in *Drosophila.* In fact, even without the new techniques enabled by the recombinant DNA revolution, the very project of studying the role of genes in *Drosophila* embryogenesis could have been counted on to substantially alter the relations between genetics and embryology. One of the most important barriers that had historically divided these two disciplines came from the absence of a shared site of investigation. Because the properties that made an organism attractive for genetical analysis were so different from those that invited embryological analysis, a material gap—constituted by the absence of an organism that had been extensively studied from both perspectives—had arisen between the disciplines alongside their methodological and conceptual differences.[27] And for anyone committed to filling this gap (as Nüsslein-Volhard and Wieschaus clearly were), the abundance of genetic information available for *Drosophila* made that embryo an obvious choice. Thus, even before the introduction of new molecular techniques, their analysis of the role of maternal-effect genes in establishing the polarity and primary axes of the *Drosophila* embryo served not only to bring these problems within the domain of genetics but also to direct the attention of geneticists to the importance of cytoplasmic factors in early development.

By the early 1980s, Nüsslein-Volhard and Wieshaus had identified virtually all the genes (both maternal-effect and zygotic) required for the formation of the basic body plan of the *Drosophila* embryo using nothing more than classical techniques of analysis. As Ashburner has written, "All this required was some standard genetics, a mutagen, and a dissecting microscope, all available in the 1930's."[28] But for the next stage of analysis, for working out just how these genes were involved in the morphogenesis of the *Drosophila* embryo, the new molecular techniques for direct genetic manipulation were indispensable: not only did they give rise to a new form of genetic experimentation but, perhaps even more importantly, they enabled the development of molecular tags for specific genes and gene products, which, in turn, make it possible to visually track the emergence and distribution of these products. And within a few short years, a clear picture began to emerge of the basic steps by which an apparently homogeneous egg is transformed—even before cell division (which in *Drosophila* does not begin until after thirteen rounds of nuclear division)—into an embryonic structure laid out along two different axes: anterior-posterior (head to tail), and dorsal-ventral (front to back). The specification of structure along the anterior-posterior axis, described by Driever and Nüsslein-Volhard, proved to be the simpler of the two processes, and also the one that provided a direct link to earlier speculations about morphogens and gradients.[29]

Here is the picture as it had emerged by 1988. The first and most important lesson was that, contrary to earlier assumptions, the cytoplasm of the *Drosophila* egg is not in fact homogeneous at any point in its development. Even before fertilization, the egg is already patterned by differential distribution of a number of specific proteins and molecules of mRNA that had been preformed (products of maternal genes) and laid down in the ovaries in which they are formed.[30] Of most immediate importance for the specification of structure in the anterior region is *bicoid* mRNA, localized at one end of the oocyte (the anterior tip).[31] After fertilization, when nuclear division commences, *bicoid* mRNA is translated into a protein (also called bicoid, but without italics) that functions as an activator of transcription for certain other genes. Since there are not yet any cell membranes, molecules of both mRNA and protein can diffuse across the egg more freely than they would otherwise be able to do, and the combined effect of diffusion, degradation, and differential translation of the mRNA into protein is an exponential gradient in the concentration of bicoid protein.

By this time, the nucleus has divided many times, and the resultant nuclei (each containing a full complement of genes) can be seen to be distributed along the inside of the cell membrane. Thus, each set of genes encounters a different concentration of bicoid, and those genes whose transcription it activates (zygotic genes) are subject to different levels of activation. As a consequence of such differential rates of transcription, gradients in the concentrations of the new mRNA molecules and proteins (products of the differential activation of the zygotic genes [for example, *hunchback*] that were directly targeted by bicoid) are now generated. Interactions among these various products in turn trigger the activation (or repression) of still other genes (for example, *giant, krüppel*), and the net effect of this cascading sequence of signals and responses is a progressively more nuanced spatial structure of bands, stripes, and segments, laid out from head to tail. Each such stripe or segment is characterized by the localization of a particular set of regulatory proteins, and in each of these, a unique combination of genes is thereby mobilized for the future development of the different body parts seen in the mature *Drosophila* (see Figure 6).

In other words, position along the anterior-posterior axis is specified by a cascade of events which, like development in general, has no absolute point of origin but which is nevertheless often said to start either with the initial localization of bicoid mRNA or with the gradient in bicoid protein to which that localization gives rise. In fact, just such a formulation is clearly suggested by the title of Driever and Nüsslein-Volhard's second paper ("The Bicoid Protein Determines Position in the *Drosophila* Embryo in a Concentration-Dependent Manner"), and even more

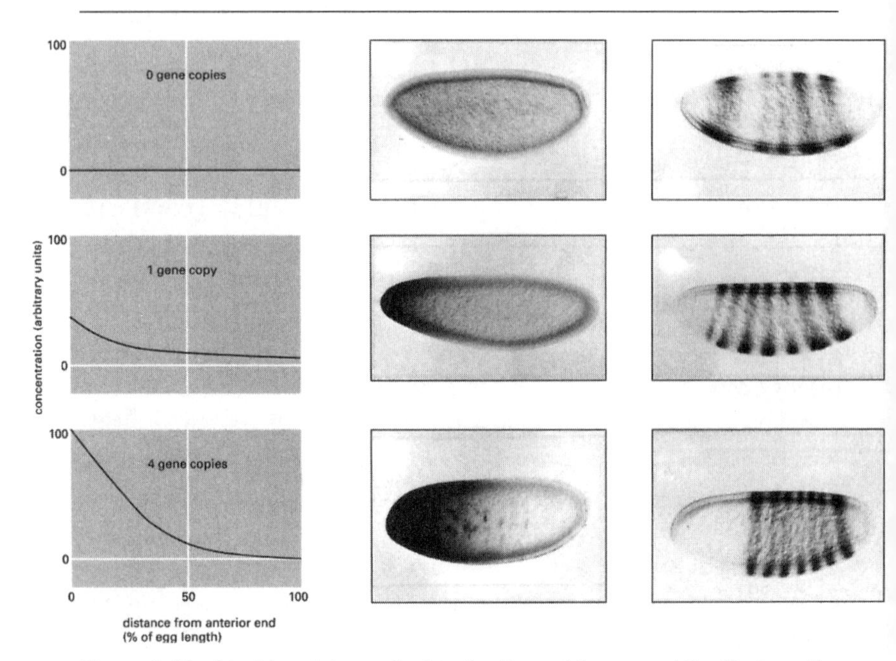

Figure 6. The bicoid protein gradient in the *Drosophila* egg and its effects on the pattern of segments. Three embryos are compared, containing zero, one, and four copies, respectively, of the normal bicoid gene. With zero dosage of bicoid, segments with an anterior character do not form; with increasing gene dosage they form progressively farther from the anterior end of the egg. (Adapted from Driever and Nüsslein-Volhard, 1988, and taken from Alberts et al., *The Molecular Biology of the Cell,* by permission from Elsevier Science)

definitively by their summary statement: "The bcd [bicoid] protein thus has the properties of a morphogen that autonomously determines position in the anterior half of the embryo."[32]

Their choice of words mattered. All of these terms—diffusion, gradient, and especially morphogen—are highly charged keywords in the annals of developmental biology, with their own histories, their own advocates, and their own epistemological resonances. Diffusion, for example, while manifestly crucial for so many physical processes, has long been regarded as ineffective for intercellular communication, and of only marginal importance for intracellular biological processes. In their recent review of the subject, Agutter and colleagues regard the persistence of diffusion theory in biology as "a relic of mechanistic materialism." They write: "The inapplicability of diffusion theory to transport processes within the living cell is well established, because of the difficulties in applying physico-chemical principles in general to the crowded, heterogeneous and highly organized interior of the cell."[33]

To many biologists, the very word raised hackles; it was a reminder of the long history of failed attempts (mostly by physicists and mathematicians) to account for biological phenomena by reducing them to simplistic cartoons that bore little if any resemblance to the clearly documented complexity of these phenomena. (One need only recall the efforts of Stéphane Leduc or of Nicolas Rashevsky.) Similarly, for a long time, the terms gradient and morphogen had had almost equally disreputable associations, for much the same reasons. The actual term morphogen had been coined by Alan Turing in 1952 for the purpose of interpreting his mathematical (reaction-diffusion) model for embryogenesis. But for most readers, the taint attached to gradients and morphogens derived primarily from the history of free-floating speculations, ungrounded by concrete experimental evidence, on the subject of morphogenetic gradients. And while there had recently been some discussion of certain small (and hence freely diffusing) molecules acting as morphogenetic signals, the role of these molecules in the regulation of specific sets of genes remained obscure.[34] Here, however, all three terms—diffusion, gradient, and morphogen—appear in the context of an impeccable demonstration of experimentally identifiable molecular processes that could be directly tied to gene regulation. This work was particularly important for the rehabilitation of morphogens. A simple search of Medline shows that use of that term as a keyword in the relevant literature took a three-fold jump between 1988 and 1989.[35] After 1988, morphogens no longer needed be thought of as a merely hypothetical construct, for now a molecule functioning with identifiable genetic specificity had been shown to play a clear and unambiguous role in the development of an animal's form, and even if not yet actually called a morphogen, it was said to have the properties of just such an entity.[36]

For Wolpert, the very words Driever and Nüsslein-Volhard chose to use was a sign of vindication. Indeed, even though no reference to his work appears in these papers, it would have been difficult for him not to find confirmation of the concept of positional information in the portrait of *Drosophila* embryogenesis they presented. Furthermore, here was a portrait that appeared to dissolve the last remnants of a divide between a purely molecular genetic approach to the problem of pattern formation and the theoretical approach he had earlier advocated which had depended on viewing these problems in their own right, and on identifying the "general universal, elegant and simple" principles that such an independent perspective permits one to recognize. Driever and Nüsslein-Volhard's work provided an obvious bridge between these two different perspectives and, at the same time, offered the promise of new disciplinary alliances.

One year later, Wolpert outlines the synthesis he envisions in a paper entitled "Positional Information Revisited." Beginning with an only slightly modified restatement of his original argument, he writes:

> Positional information provides both a conceptual framework for thinking about pattern formation and also suggests possible mechanisms. The basic idea . . . is that there is a cell parameter, positional value, which is related to a cell's position in the developing system. It is as if there is a coordinate system with respect to which the cells have their position specified. The cells then interpret their posi-

> tional value by differentiating in a particular way. This differentiation may involve developing as a particular cell type or state, or it might involve changes in growth or motility.
>
> Among the attractive features of positional information is that it provides a unifying concept for understanding the development and regulation of a variety of patterns.[37]

Put in these terms, the obvious next question is: How is such a coordinate system specified? "For a one-dimensional system," he explains, "all the necessary features can be provided by a monotonic decrease in the concentration of a chemical—a morphogen—which could be set up with a localized source or by reaction diffusion" (p. 4). And for just such a mechanism, *Drosophila* now provides his best example:

> By far the clearest demonstration of a positional signal in a developing system—clear in the sense that it can be directly visualized rather than being inferred from other properties—is in the insect egg. The gradient is in the protein coded for by the *bicoid* gene, which is a key gene in patterning along the anteroposterior axis (Driever and Nüsslein-Volhard, 1988). Its discovery is particularly gratifying not only because of its importance, but because it has just the anticipated distribution of a morphogen which is made at a source and both diffuses and breaks down. (p. 5)

Some reframing of Driever and Nüsslein-Volhard's argument was required to adapt it to fit so neatly into Wolpert's framework, but not so much as to be implausible. The picture they had provided was close enough to count as powerful confirmation.

Yet, for all the confirmation this example provided, it remains just a single case, and the concept of positional information was intended to be far more encompassing—in fact, the original hope was that it would provide a universal framework. Accordingly, Wolpert observes that it would be a mistake to think of simple diffusion as the only mechanism by which a framework of this kind could be established: "Such mechanisms could, for example, be provided by a progress zone model in which the gradient is generated as a group of cells grow, or by cell-to-cell interactions. Furthermore, the gradient of a chemical concentration is, itself, only a special case of the yet more general case." Position might be recorded by a gradient in a cell parameter of any kind, specified either internally or externally. For example, "positional value could be represented by a set of genes—additional genes being activated with increase in distance in a simple additive manner, such as 1, 12, 123" (p. 4).[38]

Within so expansive a framework, it becomes a simple matter to accommodate an immense variety of phenomena, identified by an equally disparate assortment of methodologies ranging from the newest procedures of molecular analysis to the "cut and paste" techniques of classical embryology, all under a single umbrella. These include not only early axis development in *Drosophila* but also segmentation, bristle alignment, wing patterns, and compartment boundaries in adult insects; polarity in hydra and the chick limb bud; patterning of the early frog embryo; neural tube development in vertebrates; and perhaps even the specification of the vertebral column in mice. Thus expanded, the concept of PI is sufficiently general to allow assimilation of the most spectacular reports from the new molecular developmental biology with the accumulated wisdom of classical embryology, of regeneration studies, and even of theoretical and mathematical biology.

Indeed, so general has it become, so vast the list of phenomena it now encompasses, that Wolpert himself is obliged to ask, "What would *not* be regarded as a positional system?" "There is a weak sense," he confesses, "in which the idea of positional information is used to refer to differences between cells in a developing system." But his own claim is both more specific (it does not, for example, include the specification of position by local interactions between neighboring cells) and, at the same time, stronger:

"Positional information is about graded properties and it is in this strong sense, with its implications for a coordinate system, that is considered here."

To be sure, adhering to this strong sense of the term means that something of the original claim to universality has to be given up. ("This ambitious expectation has not been quite fulfilled.") But in the recent findings from molecular biology he sees a new source of hope. In particular, the demonstration of extensive conservation across diverse phyla in the structure of certain genes that are known to be important for development (homeobox genes) and of certain proteins that are known to play a crucial role in the transmission and reception of developmental signals leads Wolpert to conclude: "Perhaps we will find that there is a common language, just different dialects."[39] It might be noted however that this common language, in contrast to his original expectations, would be in the genetic code itself rather than in the repertoire of cellular responses.

Different Ways of Knowing?

The central question of this chapter can no longer be deferred: What now—after its various reformulations, and with the accumulation of new experimental evidence—can be said of the explanatory function (and value) of the concept of positional information? Certainly, of its basic proposition, there is no longer any doubt. The cells of a developing embryo do have ways of knowing where they are—both in relation to other nearby cells and even in relation to the embryo as a whole. But was there ever any real doubt about this apparently so self-evident fact of life? In fairness, I believe it would be more accurate to say that it was the discomfort and sense of scientific impotence evoked by this obvious fact, rather than doubt, that were responsible for its being neglected for so many years. The primary novelty today is that, largely because of the new techniques that have become available, the attention of many molecular biologists has turned to this important aspect of development. And with that turn has come an explosion of detailed information about the temporally and spatially specific interactions—between nucleotide sequences and protein regulators of mRNA synthesis, between proteins and RNA, and between and among different proteins—that make up the microdynamics of developmental processes. The sheer abundance of information that has by now been amassed is as overwhelming as it is impressive, and especially so in its naked form. Thus, the challenge that confronts molecular developmental biologists today is that of integrating all this detail into a coherent developmental narrative. And here, as a way of naming this challenge, is where the most common use for PI is now to be found.

The concept of PI provides at least the form of an answer to the question of how cells "know" where they are, and especially in what Wolpert refers to as its "weak sense," it does so in a way that can scarcely be controverted: they know by virtue of their positional information. And until one attaches some particular mechanism (or class of mechanisms) to this term, such a claim is about as universal as could be hoped for. But an obvious price must be paid for universality, especially when so easily purchased: just what moves this answer beyond the realm of tautology remains entirely obscure. Without the inclusion of some description of how they know, what has now been added that would not already have been contained in a simple declarative rephrasing of the original question (that is, "cells 'know' where they are")? Wolpert is surely correct in calling this the "weak sense" of the idea of PI.

Still, weakness is not by itself a barrier to use, and indeed the term often seems to be employed in just this sense, that is, as a handy synonym for the effect of position determination (for the fact that cells "know" where they are) and where the mechanism that provides the necessary "information" either awaits specification or has been elsewhere described. Thus, when Nüsslein-Volhard and her colleagues describe the pathway of interactions between genes and proteins that leads to the specification of the ventral regions of the zebrafish embryo, they write, "This pathway provides ventral positional information," and the value of their work is clearly understood to lie in their analysis of the pathway that provides the requisite specification or "information."[40] But if weakness is not a barrier to use of the term, neither is it an incentive. Why then has it become so popular? One part of the answer is obvious: PI is certainly a more convenient expression than "cells 'knowing' where they are"; it avoids the attribution of knowledge to cells; and it represents the phenomenon in terms of information.[41] The other part of the answer, and the part that may be of more immediate relevance, is to be found in the slippage that comes as an inevitable consequence of the use of the same term in Wolpert's strong sense—that is, in the sense that clearly does carry the implication of an explanation or mechanism.

We are accordingly obliged to ask: How strong is the strong sense of PI, and what is the source of *its* strength? And here too we find a trade-off between generality and specificity. In Wolpert's most general formulation of the concept in its strong sense, "positional information is about graded properties" measured with reference to "a coordinate system."[42] Any property of a cell or its environment may count, and gradation may be observed in a correspondingly elastic sense (it may refer to the concentration of some chemical either internal or external to a cell, the time a cell spends in a particular state, the number of homeotic genes that have been activated, and so on). The importance of "a coordinate system" is that it offers the possibility of a scalar measurement (such as, for example, distance from a boundary). In other words, here the definition of the concept has been abstracted away from any particular instantiation (either material or mathematical), and by permitting it to encompass so many different phenomena, that very quality accounts for its generality. To Wolpert, abstractness itself seems to carry value, but to many others (especially to many biologists), want of specificity suggests only a lack of content.[43] Complaints of this sort can be somewhat appeased by recourse to yet a third sense of the term, namely, the original (though more circumscribed) notion of a gradient in the distribution of a chemical morphogen—a gradient that is maintained by diffusion of this chemical from its source at one end of the developing embryo coupled either with simple degradation or with some sort of sink at the other end.

Indeed, the term finds its most concrete applicability in this original sense, and its applicability comes in two very different guises. First, it continues to be useful in the construction of models for both describing and guiding a number of classical grafting (cut and paste) experiments (of the kind that Wolpert and colleagues performed on hydra; that Susan Bryant and others performed on the chick limb bud; and that Jonathan Slack, Peter Lawrence, and others performed on epidermal patterns in insects).[44] Second, it can be used to re-describe (even if not to guide) the molecular analysis of axis development in *Drosophila,* especially as reported by Driever and Nüsslein-Volhard.[45] But here, we yet again find the tension between abstraction and specificity at work. Lawrence, an advocate of PI, acknowledges that such "gradient theories" are "irritatingly abstract," and he expresses the hope that molecular studies of the relevant genes will bring these theories "to a more molecular and concrete state."[46] The difficulty is that, as they do so, these studies also undermine the value of such theories in guiding experimental work, effectively reducing their role to that of re-description—another name of an effect for which both the explication and explanation lie elsewhere.

And once again, we need to ask: What's in a name?

PART THREE

Machines: Understanding Development with Computers, Recombinant DNA, and Molecular Imaging

Everyone recognizes that scientific understanding depends on the techniques available for analysis. But the very meaning of understanding also depends on available techniques, albeit less evidently so. Both what counts as knowledge and what we mean by knowing depend on the kinds of data we are able to acquire, on the ways in which those data are gathered, and on the forms in which they are represented. Usually, however, we become aware of this dependence only in times of change, when new techniques noticeably alter our styles of knowing.

This is our situation today. In Part Three, I argue that technological advances over the last quarter of the twentieth century have already brought about discernible shifts in both the meanings and goals of explanation in developmental biology. The initial impetus for these changes came from recombinant DNA research and from new techniques of imaging living cells. But to understand the long-range impact of these new experimental techniques, we need also to consider the effects that the introduction of powerful new computers is having on biological science. For it is the computer—as a machine for processing data, for solving equations, for modeling the phenomena of interest, for refining and reconstructing visual images, and for providing four-dimensional representations of both experimental and theoretical findings to the general reader—that gives these transformations their particular shape.

Techniques of recombinant DNA have made it possible to micro-manipulate genetic sequences in ways that permit optical probes to be introduced into the interior of the living cell. The first of these transformed sequences ("reporter genes") allowed researchers to directly observe the temporally and spatially specific transcription of particular stretches of DNA. But more recent advances permit them to visually track the synthesis and subsequent activity of individual proteins as well. Consequently, molecular biologists' attention has now begun to shift to cellular mechanisms operating beyond the level of the gene. Especially dramatic is the recent discovery of a naturally occurring green fluorescent protein (GFP). When combined with new photometric detectors, charge-coupled device cameras, and computer-enhanced (confocal) microscopy, GFP labeling permits investigators to watch a protein bind its substrate or its partners in real time.

New four-dimensional computer imaging has enhanced the microscopist's perception even further. For the first time in the history of biological research, molecular processes occurring within the living cell have become visually accessible, and researchers can directly observe both the structure and operation of many of the key motors of development operating on the molecular, cytoskeletal, and cellular level.[1] Furthermore, video clips accessible through the Internet and CD-ROMs have allowed people far removed from the sites of laboratory investigations to watch the intra- and intercellular dynamics of the living embryo unfold. CD-ROMs and the Internet, I suggest, give new meaning to Shapin and Schaffer's notion of "virtual witnessing."[2]

Chapter 7, "The Visual Culture of Molecular Embryology," is concerned with the epistemic implications of these developments. The advent of technologies that allow researchers to observe directly (and others to observe indirectly) many of the processes of embryological development that occur within and between living cells constitutes a substantial discontinuity in the history of this field. To an extent that biologists could not have anticipated fifty years ago, the barriers occluding the domain that has historically been figured as Nature's innermost recess have begun to crumble. Ironically, the very effectiveness of these new technologies has contributed to the revival of a tradition that had been long-standing among students of biology—a tradition in which seeing (and, even more, watching—or seeing-in-time) is considered as both the most reliable source of knowledge and the indispensable basis of understanding. In this tradition, so widely associated with the epistemological ethos of natural history in the eighteenth and nineteenth centuries, direct observation assumes primacy; and theory—regarded as a form of speculation—is resorted to only when observation fails. As observation techniques extend into new realms, we find a revival of that ethos, and with it the return of a kind of natural history. Today, however, researchers watch the behavior of tagged molecules rather than tagged organisms.

If new visual technologies are bringing about one kind of epistemic mutation, the explosion of data they produce is making for another. I refer to the new role emerging for mathematical and computational models in experimental biology—the subject of Chapter 8. Here, too, recombinant DNA techniques have been of prime importance. The sheer quantity of data resulting from these procedures strains traditional modes of analysis—and understanding—beyond their capacity. Because many of the new experimental results can no longer be interpreted by "eyeballing" the data, and because of the increasing difficulty of accommodating them in the verbal formulations employed in the past, biologists recognize more and more the need for new modes of analysis. Some are turning to mathematical models for assistance;

others, to computational models of a kind that may or may not be mathematical in the usual sense of that term. Indeed, the computer is fast becoming an indispensable tool for most experimental biologists—not just for processing the vast quantities of data they are accumulating but for making sense of it as well.

Need is undoubtedly the major impetus for the increasingly favorable reception of mathematical and computational modeling. But presentation is another factor. New experimental results can now be represented in a format that is accessible and persuasive to an audience of experimental biologists who may be unable to follow the underlying technical analysis. Credit for this belongs not so much to the computer's computational powers as to its stunning capacities for representing those computations in visual forms that resemble familiar depictions of experimental results. Some years ago, W. Daniel Hillis remarked that "Biologists are biologists because they love living things. A computation is not alive."[3] Increasingly, however, computations are starting to look alive, and accordingly biologists are beginning to lose their traditional antipathy to mathematical and computational modeling.

Which brings us to Chapter 9, "Synthetic Biology Redux—Computer Simulation and Artificial Life." Here, I turn to the newly emerging conjunctions between simulation and construction, and to the impetus these provide for a new synthetic biology. I begin by juxtaposing Christopher Langton's program of "artificial life" with the efforts of Stéphane Leduc to establish a synthetic biology in the beginning of the twentieth century (discussed in Chapter 1). In what ways, I ask, do the possibilities of computer simulation provide new legitimation for fictionalism in science?[4] How do they add power (and perhaps even vitality) to the domain of synthetic construction?

With new developments in biological computation, the boundary between organisms and computers is becoming ever more porous. These developments are also changing our criteria for explanation, in ways that progressively blur the boundary between the sciences of organisms and computers. Where Chapter 7 focuses on epistemic mutations induced by new machines for seeing, Chapter 9 focuses on the mutations induced by new machines for doing—in particular by new technologies in biological computation and computational biology. Computer science is of course generally regarded as an engineering science, and as such it is explicitly geared to the production of practical effects. Biology, on the other hand, is a natural science, and hence one in which pragmatic goals are assumed to be at best subsidiary.[5] Today, however, as computers and computations behave and look more and more like organisms, and as organisms are likened more and more to computers and computation, the gap between the engineering sciences of computation and the natural sciences of molecular engineering seems to be closing fast. To be sure, contemporary molecular biologists still strenuously resist the most radical claims of artificial life in which digital or robotic constructions are assimilated with the organisms they study (that is, with life-as-we-know-it). Nevertheless, they live and work in a world in which what counts as an explanation has become more and more difficult to distinguish from what counts as a recipe for construction.

Yet I cannot imagine this being the last word in making sense of life. As our technical and scientific opportunities evolve, so too do our needs, and inevitably so. Explanations of development satisfying researchers at the dawn of the twenty-first century will surely not satisfy their descendants at the end of this century, any more so than Leduc's attempts from the beginning of the last century are able to satisfy us today. Nor are they likely to converge on a single account. Surely, what counts as a satisfying explanation will remain as multiple and as resistant to unity as will our needs.

CHAPTER SEVEN

The Visual Culture of Molecular Embryology

She examined the photograph of the brain cell again. "You actually know your way around one of these?"
"Not completely." He waved her chart before him casually; it was, after all, for him only so much paper. "But we're starting to get a handle on it."
"And once you've gotten it? Then what?"
He stared at her . . . "With the brain," he said, making an effort at patience, "there isn't a then-what. There's never a then-what . . . The brain's like the universe. It's inexhaustible. Your curiosity can't ever possibly be satisfied."

Robert Cohen, *Inspired Sleep,* 2001

Let us not seek for something behind the phenomena—they themselves are the theory.

Goethe, *Scientific Studies,* 1817

ne of the first lessons students are taught about the scientific method is of the importance of distinguishing between evidence and explanation. Evidence, at least in the root sense of the term, is that which can be seen, and whether or not one takes observation to be theory-laden, the relation of evidence to explanation is generally construed as either confirming or disconfirming—that is, evidence is either for or against a proposed theory or explanation. Conversely, an explanation is either able or unable to account for the evidence. The failure to make this distinction is widely taken as a failure to understand one of the most basic principles of scientific reasoning.

But need it be so? Does not evidence, especially in the sense of that which has been made observable, under some circumstances also have standing on its own—not merely *for* (or against) a theory, argument, or hypothesis but also *of* a phenomenon that is neither more nor other than itself? And are there not circumstances in contemporary scientific practice when the mere observation of a phenomenon is itself so satisfying and so compelling that no further explanation seems to be required? When seeing is itself a kind of understanding, and not just in the colloquial sense of "I see"?

There is of course a long tradition in which understanding is itself taken to be a kind of seeing. The colloquialism "I see" is hardly innocent, for it indicates the depth with which the meaning we give to understanding has been bound up with seeing, and the difficulty of speaking—or for that matter, of thinking—about understanding without invoking the metaphor of vision. The aim of science is to discover Nature's secrets, to see her unveiled. To explain is to make things "clear and evident," to illuminate and enlighten. We have understood when we have seen with the mind's eye. The visual metaphor for knowledge is everywhere.[1] And as is the way with metaphor, it simultaneously reflects and enforces a dynamic interdependence between mind and eye too complex to permit disentangling, and too embedded in our cognitive apparatus to do without.[2]

Consider, for example, the many ways in which we depend upon visual representation for understanding. Edward Tufte refers to the use of graphic imagery for explanatory purposes as "visual explanation," and he argues "that clarity and excellence in thinking is very much like clarity and excellence in the display of data. When principles of design replicate principles of thought, the act of arranging information becomes an act of insight . . . By extending the visual capacities of paper, video, and computer screen, we are able to extend the depth of our own knowledge and experience."[3]

Similarly, the mind's eye and the actual eye are also (and equally) conjoined by the ways in which seeing itself depends on prior understanding, by the dependence of observation on the assumptions (and theories) we bring with us.[4] We have learned that there is no such thing as naïve seeing. These two different ways in which seeing and understanding are entwined are familiar concerns in the history and philosophy of science. But we might also ask an even simpler question, one that bears on a yet more immediate way in which evidence and explanation, the *what* and the *how,* may be entangled: Is clarity in thinking always and necessarily of higher epistemological value than clarity in seeing? Indeed, is it always possible even to distinguish these two kinds of seeing from each other? Are there not circumstances in which seeing is itself a kind of knowing? When evidence achieves the purity that Lorraine Daston attributes to seventeenth-century mirabilia, "unequivocal in its interpretation and irresistible in its persuasive power"?[5]

Daston's investigations into the historical specificity of such kindred concepts as facts, proof, and objectivity are very much to the point here. For example, she takes current understandings of facts as "brute," robust, and inert, as "the mercenary soldiers of argument" and hence as belonging to a category definitively other than and clearly distinguishable from that of "evidence" and asks: Was it always so? (pp. 243–244). For if it was not, then another question immediately presses itself upon us: How did our current conceptions of neutral facts and enlisted evidence, and the distinction between them, come to be? In much the same spirit, it seems to me useful to inquire into the equally canonical disjunction between evidence and explanation, and ask not only "Was it always so?" but also, even now, "Is it everywhere so?"

For if we can identify scientific practices in which the stuff of observation is taken as unequivocal and irresistible, then our understanding of the relation of evidence to explanation ought also to be grist for Daston's mill, and possibly in a more extended sense of her "historical epistemology." Such practices may be far less exceptional than is normally assumed. Indeed, I suggest they seem particularly likely to be found in the biological sciences, indicating disciplinary as well as historical heterogeneity. If I am right, we need to ask how epistemic categories relate to the questions, practices, technologies, and *mentalités* characteristic not only of particular times but also of particular scientific endeavors at a given time. Such a project, even if confined to the relation between evidence and explanation (or between seeing and knowing), is of course vast, and the remarks that follow are intended merely as preliminary notes. They are prompted in the first instance by the particular significance of visual evidence throughout the history of embryology and, second, by specific recent mutations in the technical practices of molecular embryology associated with an entirely new order of visual access to the living embryo.

Nature's "Innermost Closet"

It is difficult to talk about things that are obvious, about images that have become so familiar as to be effectively invisible. For example, when feminist scholars first began to call attention to metaphors of gender in the language of science, one of the principal obstacles they encountered lay in the very ubiquity of such metaphors. The most obvious figure was of course Mother Nature, her secrets hidden from view, simultaneously provoking and resisting the penetrating gaze of science, but here was a metaphor so commonplace as to have become effectively unnoticeable. Even when noticed, its significance was often discounted on the grounds of its being a "dead" metaphor and hence devoid of force. But metaphors are dead only because we cease to notice them, because we are no longer conscious of their effects on our perception. It might even be argued that dead metaphors are the most forceful of all, just because their mode of operation is beyond the realm of consciousness, effectively screened by their very banality. In any case, all dead metaphors were once alive.

Certainly, the figure of the maternal womb as the harbor of primal secrets was once very much alive, not only in historical time but in the lives of all of us as inquirers, as seekers of knowledge, and it left its trace. In the early history of science, the mystery of embryonic life provided a readily accessible image for representing Nature's ultimate secret; it could stand for the unknown precisely by virtue of being so deeply hidden, so fully sequestered beyond the range of human vision. When Henry Oldenburg described the aim of science as "penetrat[ing] from Nature's antechamber to her innermost closet," when Anton van Leeuwenhoek wrote of the pleasures of penetrating "the arcana of nature," or when Jean Senebier claimed of hypotheses that they are "the resort of the Physicist who cannot be instructed by observation alone," turned to only because "Nature almost forces him, with her obscurity, to imagine what she insists on hiding," they were all directly or indirectly invoking the image of Nature's womb and the secrets it contained to represent the object of scientific inquiry in general.[6] That image evoked powerful resonances, simultaneously expressing and forging a link between seeing and knowing that remains in force to this day. When it was said (and is still sometimes said) that Mother Nature hides her secrets from us, the secret literally concealed within the body of the mother was never far behind. Indeed, these two kinds of secrets are so proximate and so closely intertwined in the history of modern science that the unraveling of the latter has frequently been taken as a synecdoche for the former, with embryology standing not only for the science of biology but for all of natural science.[7] Furthermore, the proximity of these two kinds of secrets evokes another image as well, that of the scientist as voyeur, peering into nature's darkest corners.

In the actual science of embryology, however, the synecdoche dissolves. There, the mystery of the growing embryo *is* the object of scientific enquiry. And with that collapse between the metaphoric and the literal, so too does the gap between seeing and knowing threaten to close. But even here, and even now, the force of the embryo as a figure for the more general object of scientific inquiry is still in evidence, especially in the continuing reference to the process by which the microscopic egg expands and develops into a human adult as "the mystery of mysteries."[8] In our inevitable egocentrism, the secret of life is in the final instance the secret of human life; and for many, this riddle, the mystery of our own origins, remains the ultimate and most compelling secret of all.

But whether in embryology or in science more generally, the question of how best to gain access to nature's secrets remained. In fact, this is the question over which the life sciences historically set themselves most decisively apart from the physical sciences. Whereas for Descartes, rational thought provided the most reliable means of knowing, by the late seventeenth century, the recalcitrance of problems like generation and development had already precipitated a strong reaction against Cartesian mechanism among life scientists, and their reaction was grounded precisely in the epistemological imperative they claimed for actual seeing. Students of living forms countered the primacy of the mind's eye with the primacy of the corporeal eye, claiming epistemological certainty for direct observation rather than for deduction, relying on instruments of sight rather than on those of calculation. The two great technical achievements of the seventeenth century, both of which were of inestimable importance in the subsequent development of modern science, were the microscope and the calculus. And in terms of this particular controversy, these two achievements—like their two famous protagonists, Robert Hooke and Isaac Newton—clearly weighed in on opposite sides.[9] Indeed, a certain parallel might be drawn between the role of the microscope in the growth of the life sciences and that of the calculus in the progress of the physical sciences.

The Biological Gaze: Seeing and Watching

The importance of observation in the history of biology can scarcely be overestimated. As N. J. Berrill reminds us, biology is, and always has been, an "eminently and inherently visual" science.[10] In fact, its particular reliance on visual evidence may shed some light on the troubled history of the role of mathematics in biological science that I discussed in the first part of this book. There I described the difficulty in terms of a tension between theory and experiment, but perhaps what has been at issue is rather a tension between imagining and seeing—that is, an opposition between what may be imagined with the help of mathematical and mechanical models and what can actually be seen with one's own eyes. For Senebier was certainly not alone in relegating theory to the status of a default option, in suggesting that we find recourse in reflection, in speculation, in seeing with the mind's eye only when actual processes are hidden from view. Almost a hundred years earlier, the microscopist Jan Swammerdam had written, "The philosophers' true knowledge consists only in the distinct idea they may have of the effects that strike their eyes."[11] For these men, as for so many of the life scientists who followed, it was in the first instance Nature's obscurity that obliges us to imagine, to make hypotheses, to depict in theoretical schemes and representations that which we cannot literally observe.[12]

Seeing, then, was the thing, and the microscope was its enabling instrument. Or perhaps seeing is not quite the right word. For what these early life scientists were doing as they gazed through their microscopes might better be called watching. The microscope revealed to them the spectacle of life on a scale heretofore undreamed of, and what they saw through their lenses was the marvel of life in action. The new visual technology provided them with the means not only of observing the structures of the minute organisms and animalcules they discovered but of seeing these structures as alive and moving (Leeuwenhoek called them "living atoms"), in a habitat apparently unperturbed by the inquiring eye.[13] Hooke wrote, "We have the opportunity of observing Nature . . . acting according to her usual course and way, undisturbed," and he contrasted the effect with that which is seen "when we endeavour to pry into her secrets by breaking open the doors upon her." Here, he continued, "we find her indeed at work, but put into such disorder by the violence offer'd, as it may easily be imagine'd how differing a thing we should find, if we could . . . quietly peep in at the windows, without frighting her out of her usual byas."[14] The result, as Jacques Roger put it in his classic and still unsurpassed history of life science in the Enlightenment, was that "Men found themselves surrounded by living beings."[15]

To be sure, the microscope introduced its own sources of uncertainty. How could one be sure that what one was seeing was real, and not an artifact either of the instrument or of the imagination? Furthermore, early microscopes were hard to come by, and access was severely limited. Could one trust the reports of those few who did have access? Finally, seeing through this instrument required enormous skill, of a kind that was not easily transmitted. Life scientists may have granted priority to seeing over speculating, but did the microscope in fact provide for true seeing? Such skepticism was by no means gratuitous, for the invention of the microscope proved not to be capable, in itself, of fully bridling the imagination. Perhaps nowhere are its limitations clearer than in the early history of embryology. Leeuwenhoek's sightings of fully formed animalcules in the male seed in the late seventeenth century are a familiar reminder that seeing need not be prior to believing.[16] In fact, early microscopes may have been particularly prone to error, to artifact, to fantastic projection, and this is undoubtedly part of the reason such a long time passed before confidence in their reliability took hold at large. But even after confidence had been firmly established, instances of illusory sightings recurred throughout the history of microscopy; and they clearly demonstrate both the uncertainties of vision and the impossibility of naïve seeing.

Yet despite its fallibility, it was the microscope that ultimately spelled the demise of the more baroque speculations of seventeenth- and eighteenth-century embryological theorists. The great age of biological microscopy came in the nineteenth century with the arrival of better lenses, greater magnification and resolution, improved preparation of microscopic specimens, and, above all, renewed confidence in the veridicality of the basic instrument.[17] Nineteenth-century microscopy enabled virtually all the classic observations—of eggs, sperm, fertilization, and the contours of embryonic cleavage—on which modern embryology is based. Indeed, it is difficult to imagine how the subject could have developed without it. As Berrill writes, "Much of its progress during the past two centuries has resulted from the invention of visual aids ranging from simple magnifiers to the scanning electron microscope."

This much is standard history. But given how much more than powers of resolution separates simple magnifiers from the electron microscope, it is somewhat surprising to find Berrill passing so easily from the one to the other. Indeed, between the two lay a critical advance in visual technology that was of particular importance for experimental embryology, and it came long before the arrival of the electron microscope. It too was a product of nineteenth-century advances in visual technology, and the gap it opened up reveals the role that these advances in biological microscopy also played in splitting experimental embryology off from the other life sciences, dividing biology into two domains—one of the living, and the other of the non-living.

While simple magnifiers and early microscopes extended our optical capacities into the domain of smaller and smaller life forms, culminating in the identification of the cell in the 1830s as the primary element of living matter, it was there, at the cell membrane, that microscopic studies of the living world reached the limits of their capacity. Particularly for animal cells, the cell membrane marked the edge of visibility. This impasse arose in part from the limits of the microscope's resolving power, in some cases from the opacity of the cell membrane, but more seriously by far from the fact that the interior structures of the cell are inherently colorless, translucent, and hence all but invisible. Seeing beyond the membrane of animal cells required hardening the cell, cutting it into thin slices, and immersing it in dyes and stains that would heighten the contrast of the internal structures so as to render them visible.

Technical advances in cutting, fixing, and staining were responsible for the emergence of modern cytology. But their very success also took biological science in new directions, away from the detection of miniature forms of life and toward studies of the microstructure of biological entities from which all signs of life had departed. Biologists could see more, but they could no longer watch. If the demands of cutting and fixing did not themselves disrupt the life of the cell, then those of staining did, for the most effective stains available until the second half of the twentieth century could be taken up only by non-living structures. Enormous strides were made in the identification and characterization of those intracellular structures that available stains could mark—

most notably the nuclear threads or chromosomes.[18] But cytoplasmic structures were far less accessible, in good part because they are so much less stable than the chromosomes.

As late as 1925, E. B. Wilson wrote in his classic treatise on *The Cell in Development and Heredity:* "We are driven by a hundred reasons to conclude that protoplasm has an organization that is perfectly definite, but it is one that finds visible expression in a protean variety of structures . . . The fundamental structure of protoplasm lies beyond the present limits of microscopical vision and hence still remains a matter of inference and hypothesis."[19] Because of their protean nature, seeing these structures would require not only higher microscopic resolution but new ways of seeing; above all, they would require techniques for seeing-in-time. Even if these structures could be caught sight of with microscopic vision, only seeing them in their temporal development could provide assurance that what one was seeing was in fact real, and not merely an artifact.

For Ramón y Cajal, one of the great histologists of this time,

> every advance in staining technique is something like the acquisition of a new sense directed towards the unknown. As if nature had determined to hide from our eyes the marvellous structure of its organization, the cell, the mysterious protagonist of life, is hidden obstinately in the double invisibility of smallness and homogeneity . . . The histologist can advance in the knowledge of the tissues only by impregnating or tinting them selectively with various hues which are capable of making the cells stand out energetically from an uncoloured background. In this way, the bee-hive of the cells is revealed to us unveiled; it might be said that the swarm of transparent and invisible infusorians is transformed into a flock of painted butterflies.[20]

But, as Hannah Landecker observes, "Butterflies, perhaps, but hard, dead, unmoving butterflies."[21] Her reading of the history of cell biology after Cajal is very much to the point here: "It was not a question of making a thing visible that had not been seen before . . . It was a question of making a *process* visible by seeing the thing change continuously over time" (p. 68).

Perhaps nowhere is seeing-in-time more necessary than in the study of the unfolding of living form in embryonic development, and Landecker's study includes a useful discussion of the impact of cinematography on work in this field in the early part of the twentieth century. But even with cinematography, the interior processes by which a fertilized egg gives rise to a complex organism remained invisible—as Wilson wrote, "beyond the present limits of microscopic vision." For the observation of these processes, the preparatory techniques that had been so crucial to the advances of nineteenth-century cell biology were of little help. Given the nature of their subject, it might be said that, well into the twentieth century, experimental embryologists were obliged to restrict their study to extracellular dynamics. Here they achieved some remarkable successes.

Making use of relatively low-power microscopes, of the occasional intracellular coloration provided by nature (for example, in the yellow crescent of frog and sea urchin eggs), of natural stains that would mark the surface without disrupting the life of the cell, and of observational skills developed over years of practice, they were able to track the outer processes of embryonic cleavage and follow the orientation and movement of the daughter cells in the growing embryos of a number of species.[22] But they had little access to the processes occurring beyond the cell membrane, and the possibility of observing these at the level of individual molecules was absolutely beyond their ken. Indeed, it is for just this reason that Berrill—by practice and by training a classical embryologist—exempts the molecular level when describing biology as "eminently and inherently visual."[23]

Bringing molecules within view was the great achievement of the electron microscope.[24] The advent of this new visual technology in the middle of the twentieth century—increasing powers of resolution by as much as three orders of magnitude—undoubtedly marked a triumphant advance in the history of microscopy. But from the perspective of embryology, its impact on biological science seemed only to recapitulate the history of nineteenth-century microscopy.[25] Just as with so many of the technical achievements of the earlier period, the price that had to be paid for the new powers of resolution, now extending all the way down to the molecular level, was the suspension of all biological activity.[26] Thus, it too was effectively useless for the study of cellular and developmental dynamics.

To be sure, other, and roughly contemporaneous, developments in optical microscopy were already beginning to breach the barrier posed by the cell membrane. With the phase-contrast microscope developed by Frits Zernike in 1932 (although not generally available until after the end of the war), with the use of antibodies to detect intracellular antigens beginning in 1941, with Georges Nomarski's differential interference contrast microscope built in 1952, and with Shinya Inoué's improvements in polarized light microscopy in the early 1950s, the drama of life's unfolding held new allure for cell biologists. Inoué's achievements were of particular consequence, for they were the first to bring the elusive mitotic apparatus of cell division into clear view. This work also illustrates the power of real-time viewing of living cells in adjudicating conflicts and resolving uncertainties about the reality of what has been seen. Inoué succeeded in persuading biologists that the spindle fibers and fibrils he observed were real structures—and not, as had been argued for decades, artifacts of fixation—precisely because he was able to exhibit them in "living, normally dividing cells."[27]

But no improvement in technique could extend the power of resolution of optical microscopes beyond the limit imposed by the wavelength of visible light, and despite even the substantial advances that developments in optical microscopy brought in our ability to see inside living cells, these were soon overshadowed by the dramatic increase in resolution provided by the electron microscope. The new kind of microscope extended the range of objects that could be seen down to the level of a single angstrom. Its power was hard to resist, and once again the allure of visibility quickly overtook that of vitality in the imagination of many (if not most) biologists. As Robert Allen wrote, "Following the introduction of ultramicrotomy in the 1950's, many biomedical scientists shunned the light microscope in favor of electron microscopes, which had a thousand times better resolving power. In doing so, in a sense they gave up the study of living processes, except for indirect evidence that came to them fortuitously from electron micrographs."[28]

Indeed, we might argue that the successes of electron microscopy represent the culmination of biology as a science of the non-living. Its achievements seemed only to reinforce the conviction of an essential incompatibility or impasse between visibility and vitality—or, as Niels Bohr had famously argued in 1932, between "light and life."[29] It seemed only to confirm the belief, born of nineteenth-century advances in cellular microscopy, that if one wanted to know about the most intimate details of life inside the cell, it was necessary to choose between seeing and watching—between observing the static residues of cellular life that remain after fixation and following the behavior of the living cell from without.

Remarkably, however, that impasse seems now to be giving way. Thanks to the infusion of new technologies from recombinant DNA and computer science into biological microscopy, it has become possible not only to observe in vivo the temporal dynamics of many of the intra- and intercellular processes involved in embryogenesis, but to do so down to the level of single molecules. With the help of these new techniques, molecular embryology begins to close the gap separating it from its classical precursor: it too becomes a science of vital proceedings. Advances in visual technology promise the realization of an age-old quest: they enable researchers to watch the mystery of mysteries unfold before their very eyes. And, with the help of computer recordings and reconstructions, the rest of us can participate in this astonishing spectacle as virtual witnesses.

The question such achievements raise is two-fold: How do they affect our understanding of embryogenesis, and how do they affect our understanding of understanding?

Crossing the Vital Barrier

The three principal technical advances responsible for the revitalization of embryology are closely interlinked. They are: (1) techniques for introducing molecular markers that can serve as visual probes of the cell's internal dynamics; (2) optical instrumentation for viewing those markers; and (3) techniques for processing and visually representing that information. Each of these advances bears on one aspect of the complex of activities required for biological visualization. The first concerns modes of intervening; the second, of looking; and the third, of representing.

Molecular markers can be introduced into living cells in a number of different ways: by direct injection of optically tagged proteins or other molecules into the cytoplasm, by insertion of "reporter genes" directly into the DNA using recombinant DNA techniques, or by injection into the cytoplasm of a photosensitive precursor of a molecule (a "caged molecule"), which, once activated by a pulse of laser light focused on a chosen site in the cell, can be visually tracked in its subsequent spatial and temporal dynamics.

An example of the first procedure is the microinjection of a purified protein that has been coupled to a fluorescent dye. Because the injected protein can be seen through a fluorescent microscope, its fate can be tracked visually as it is incorporated into the machinery of the growing and dividing cell. Alternatively, one might inject a fluorescent labeled antibody that is specific to the protein of interest. Optical labels in the DNA—so-called reporter genes—are site-specific sequences of DNA that attach optical labels to particular mRNA and protein molecules.[30] Some reporter sequences provide sites on the transcribed mRNA to which microinjected fluorescent dyes can bind (thus making the mRNA visible by fluorescent microscopy), while others may code directly for a naturally fluorescent protein (thereby making the protein itself visible). When inserted immediately adjacent to either the promoter or the coding sequence of a protein, one kind of reporter gene permits investigators to observe the transcription (or expression) of the adjacent sequence, while the other kind permits them to follow the appearance and subsequent behavior of the protein that is synthesized from that sequence. The third procedure described above permits the visual tracking of smaller molecules and is especially useful for the observation of intracellular signaling.

All of these techniques are obviously invasive; their usefulness for observing in vivo processes will depend on the degree to which they disrupt the normal course of cellular events and the rate at which they may compromise the life of the cell or organism. A fluorescent tag that has generated particular excitement over the last few years—in good part because its attachment to the resident proteins of a cell does not seem to interfere with the function of these proteins—is the intrinsically fluorescent GFP (green fluorescent protein) obtained from the jellyfish *Aequorea victoria*. Because of the wide range of molecules to which GFP can be attached, it has acquired the status of a "magic lantern," an almost universal tag for making proteins visible in living cells, "encompassing everything from neurotransmitter receptor proteins to tiny intermolecular binding motifs." Andrew Matus continues, "No longer must we struggle to appreciate the molecular dynamics of cellular function from diagrammatic reconstructions or by assembling single images of fixed cells in our imagination—now we can see these events directly, and through this, our understanding takes a quantum leap . . . Especially for those of us raised on immunostained images of dead cells, this sudden blossoming of molecular dynamics in vibrantly alive cells and tissues is an unexpected miracle."[31]

Moreover, GFP can be modified by site-directed mutagenesis to produce different-colored variants, thereby enabling the visual distinction of a protein tagged with one variant from a second protein tagged with another. Such color differentiation makes possible not only the tracking of individual proteins but also observation of protein–protein interactions *in situ.* In short, not only can investigators watch where proteins go but they can also see proteins "at work," "watching what [they] are doing when they get there." Michael Whitaker has likened the practice of new microscopic techniques to that of entomology, where "entomologists study the foraging behaviour of individual ants by applying different coloured paint spots to them as they leave the nest."[32]

Visual probes of the kind just described have had a dramatic impact on biological research, especially in developmental biology. They have enabled biologists to track the spatially and temporally specific dynamics of gene expression and protein function in developing embryos, and in so doing have contributed substantially to our current understanding of these dynamics. But even more startling is the visual impact of these probes, which—when coupled with independently developed visual technologies—permit four-dimensional viewing of molecular and cellular mechanisms of development. Of particular importance has been the use of video and confocal microscopy, in conjunction with computer processing and with increasingly sophisticated methods for keeping embryo cultures alive under their scrutiny.[33]

Video-enhanced contrast microscopy was first introduced into cell biology in 1981 when Shinya Inoué, Robert D. Allen, and colleagues found that the analog controls of a video camera could be used both to amplify the light intensity of images too weak to be seen with the human eye and to enhance the contrast of these images.[34] Coupled with digital processing, the combination of intensification and enhancement increased by an order of magnitude the range of objects that could be visually detected. Finally, the speed of television cameras added a parallel increase in temporal resolution. With such improvements, video microscopy brought within view many in vivo structures and processes that had not previously been visible. Initially, the new technique was employed primarily to elucidate the dynamics of microtubule assembly and disassembly in the processes of cell division and motility. However, with the subsequent developments in confocal laser scanning microscopy (CLSM) and computer reconstruction, the capacity of video microscopy has since been extended to permit the viewing of the structure and dynamics of a far greater range of intra- and intercellular activity.

The main limitation of video microscopy lay in the thinness of the tissue preparations required and hence the impossibility of imaging thick specimens. And it is just here that the most dramatic achievements of confocal microscopy lie. The confocal aperture (or pinhole) focuses the beam precisely so that disturbances of light reflected from above and below the plane of focus are removed; the use of laser light adds the intensity needed to make the specimen visible; and a scanning (or raster) device permits the generation of a two-dimensional image of a small area of confocal spots (an optical section). As the laser scans the specimen, the analog light signal is picked up by a photomultiplier and converted into a digital signal. Because the plane of focus can be selected and moved up and down by a computer-controlled motor, such a system allows for the examination of specimens of considerably greater thickness than is possible with an ordinary light microscope. Because the digital image can be processed to amplify differences in light intensity too subtle to be seen by the human eye, details not otherwise visible can be rendered visible. Moreover, because computer processing can reduce noise (resulting, for example, from imperfections in the optical components or from the spurious effects of double refraction) by the use of averaging techniques, still further refinement of the image is possible. Finally, computer software has been designed to stack all the information that has been gathered and reconstruct it to produce a three-dimensional image of the specimen.[35]

The human observer reenters the scene only at the end of this series. And here another problem arises, for there is of course also such a thing as seeing too much. The expanding repertoire of molecular probes and the increasing power of imaging instruments have made so much visual information accessible that its very magnitude can be confounding. How much detail can the human eye process before becoming overwhelmed? And how can so many data, especially when seen in their temporal unfolding, be shared with and relayed to distant colleagues? If the role of the computer in the generation of these problems has been central, its role in their resolution has become even more so. Indeed, it is in dealing with these problems that the computer has become indispensable for contemporary biological research. To overcome the limits of both perception and communication, researchers rely on computer imaging software to select and highlight data of particular interest, to display the data in perceptually manageable forms, and to record them in formats that can be easily transmitted to others. Even though the image actually seen by human viewers (whether by those looking through the microscope or by those who have access to it only in its recorded form) is the result of an elaborate system of computer processing and reconstruction, it can be so compelling that viewers have the impression they are looking directly at the specimen in its naked reality.

The illusion of veridicality is made even more compelling by the presentation of these three-dimensional images in time. This is a particularly simple matter for images taken from a living cell or embryo, for they actually appear sequentially, and the microscopist at least has firsthand access to the temporal processes unfolding before his or her eye. But even when the original data have been gathered from a non-living system, a four-dimensional image can be reconstructed without difficulty—provided, that is, a population of more or less identical specimens is available for fixing at successive stages of development. Indeed, the remote viewer who is not actually at the microscope often cannot tell whether the four-dimensional image has been produced from observations on a living or a non-living system. Nevertheless, time-lapsed films, reconstructed from the confocal images of fixed, successively staged, systems, have proven of immense value in cell and developmental biology—especially in the study of cell division and cytoskeletal development.

The work of Victoria Foe and Garrett Odell provides a good example of the value of such an approach for studying the morphogenetic development of early embryos. Foe and her colleagues have recently produced a series of stunning movies of the dynamics of filamentous actin, myosin, and microtubules in syncytial (pre-cellular) *Drosophila* embryos that clearly show the temporal and spatial changes in the organization of these cellular components over the course of the cell cycle.[36] By studying the effect of drug injection experiments on the concentration and kinematics of actin and myosin filaments and, in particular, on their specific affinity for the cellular cortex, they were able to hypothesize specific chemical and mechanical interactions capable of accounting for early morphogenesis in *Drosophila* (and possibly in other organisms as well) in concrete physical terms.

Studies such as these depend critically on the remarkable improvements in visual instrumentation that have emerged over the last thirty years, and especially on the power, speed, and accuracy of the new imaging technology. They do not, however, depend on being able to employ this technology on living embryos. The images one sees may look alive, but just like the cinematography of the earlier part of the twentieth century, they are produced from images of dead embryos. The obvious question arises: Does it really matter? And if so, why?

As I see it, living specimens offer at least two critical advantages for microscopic analysis. First, because the time required for computer processing is so much less than that required for the preparation of fixed embryos, observation of images taken from living specimens permits detection of processes occurring too rapidly to be captured in time-lapse photography of successively staged embryos. Second, and perhaps even more important, is the fact that they permit the tracking of individual molecules and cells. There is no way of tagging single molecules so that one can recognize the same molecule in two different preparations, and only in rare cases is it possible to label the individual cells of an organism in ways that permit one to identify a cell in one slide as the same cell appearing in a similar position in another slide.[37] To distinguish individual behavior from the average behavior of a population, one needs to be able to track a particular molecule or cell in real time.[38]

Confocal microscopy on living systems is not easy, and the problems that arise come from biology rather than from engineering. These have to do with the difficulty of maintaining cells or embryos in a healthy state under the conditions of observation. For the study of living systems, the utility of this technology has depended on the identification both of nondisruptive visual markers and of methods for providing the protection from photodynamic damage that is required to maintain normal function. In fact, it was principally improvements in cell culture technique that inaugurated the use of the confocal microscope for in vivo observations of cellular and developmental processes.[39] Over the last decade, substantial strides have been made in both instrumentation and biological technique, and together these have made confocal microscopy a prominent tool for studying the activity of individual cells and molecules. One effect has been an increasing focus on molecular and cellular individuality. As Roger Tsien puts it, "When you can see their individual biochemical signals, you find that different cells are often very individualistic, almost like wild animals, or people."[40]

The same theme recurs in Robert Service's description of the use of confocal microscopy for "Watching DNA at Work." Despite the dramatic progress that has been made in sequencing DNA, Service notes that "in all these studies, researchers examine the collective, herdlike behavior of many thousands of copies of particular DNA fragments. Now . . . a quiet revolution is under way . . . Just as an ecologist uses radio collars to track the movements of individual animals, these researchers are using tools such as lasers and magnets to gain a wealth of new insights into how DNA twists, turns and stretches."[41] For studies more directly pertaining to embryonic development, however, I turn to the work of Scott Fraser and his colleagues, who employ CLSM in the study of both the intra- and intercellular dynamics of development in intact embryos. This work clearly illustrates the value of this new technology not only for discovering new facts and for stimulating interest in molecular mechanics but also for generating substantive shifts in our overall perspective on development.

In one study, Miller and his colleagues were able to observe the dynamics of thin filopodia during sea urchin gastrulation. Gastrulation is the process in which the primary embryonic architecture is established, and it involves a dramatic rearrangement of the constituent cells.[42] How is this rearrangement carried out? How do the cells "know" where to go? Fine threadlike structures emanating from mesenchyme cells had earlier been glimpsed with time-lapse and differential interference microscopy, but their function had remained in considerable doubt. Gustafson and Wolpert had speculated that these structures might act as "sensors" that somehow pick up patterning information from the ectoderm through the diffusion of molecules in the substrate, and then move the cells accordingly.[43]

Now, confocal microscopy has definitively established both the presence and exceedingly rapid movement of long (over 80 microns) and very thin (.02–.04 microns in diameter) filopodia extending not only from mesenchyme cells but also from ectodermal cells. These structures provide a means for intercellular communication over a distance of several cell diameters, but molecular diffusion appears to play little if any role. Rather, the thin filopodia enable each mesenchyme cell to make contact with more than 50 ectodermal cells through direct receptor-ligand interactions between the cell membranes of the extending filopodia.[44] By observing the behavior of these structures under a variety of experimental perturbations, Miller et al. were able to show that they are only indirectly responsible for cellular motion: the cell's motors are now thought to be activated only after the information collected by the filopodia is relayed back to the main body of the cell.

In a more extended series of studies, Marianne Bonner-Fraser, Scott Fraser, and Paul Kulesa have employed confocal microscopy to explore neural crest cell migration in developing chick embryos, and here, too, their findings have been new and surprising—in this case, concerning a phenomenon that had been extensively studied by both classical and molecular embryologists. But where earlier studies had been based on in vitro studies of cell behavior, the new technology provided the means for tracking the movements of these cells in their normal environment. Following the trajectories of individual (fluorescent-labeled) cells in intact embryos throughout the course of their migration enabled them to show that these trajectories are inherently unpredictable. "The unpredictable cell trajectories, the mixing of neural crest cells between adjoining rhombomeres, and the diversity in cell migration behavior within any particular region," write Kulesa and Fraser, "imply that no single mechanism guides migration."[45] Subsequent work from the same laboratory argues strongly that the directionality of cell motion is not, as had been previously thought, pre-specified, "programmed" by the patterns of gene expression established in the neural tube, but rather that it arises dynamically from ongoing interactions with other cells and with their local environments. Indeed, the current view of these authors is that observed patterns of gene expression seem to follow from rather than to determine the cell's targeted destination.[46]

Implications

Confocal microscopy is only one of the new visual technologies available to molecular embryologists. And for all its successes, CLSM has distinct limits. Even today, visibility in intact embryos remains confined to a few hundred microns, and the best techniques of embryo culture provide a window of only a day or two. Consequently, a number of researchers have begun to look to other techniques for extending their visual access. Some are turning to magnetic resonance imagery (MRI), until now used primarily for medical imaging, as a less invasive method for "looking deeper [and longer] into vertebrate development."[47] Because MRI is based on nuclear magnetic resonance signals rather than on optical radiation, it is simultaneously more penetrating and less disruptive than confocal microscopy. Thus, when used in conjunction with appropriate magnetic contrast agents, it can generate in-depth images of living embryos over a period of several days. Other researchers have turned to two-photon excitation microscopy as a method that drastically reduces photodamage, or to atomic force microscopy for direct observation of conformational change in proteins.[48] New techniques abound, and they are developing at an accelerating pace—spurred in large part by the growing interest, among an increasing number of researchers, in the opportunities and new perspectives generated by already existing technology.

I began this chapter by raising a number of questions about the historical relations between seeing and knowing in scientific explanation, and I suggested the existence of a long tradition, especially pronounced among life scientists, in which direct observation is granted priority over other ways of knowing. On the most general level, we might say that the new technologies have encouraged a revival of this tradition. Without question, they have brought about a renaissance of microscopy in the biological sciences. And with that renaissance has come a renewed emphasis on the epistemological importance of what is still referred to as "direct observation," even with the increasing reliance on computer reconstruction for making optical data visible. Indeed, even images that are not viewed through a microscope (and in some cases could not in fact be seen through a microscope) are taken to provide visual evidence.[49] Seeing is believing, or so alleges the title of an increasing number of articles.[50]

Furthermore, seeing is for many also a kind of understanding, satisfying in and of itself. To observe the molecular motors that drive embryonic development, to watch them pushing and pulling the various components of the cell into shape, regulating both the composition and conformation of particular proteins and transporting these proteins to temporally and spatially specific sites of activation, is to see the machinery of life in action and hence to go a long way toward understanding how a single cell transforms itself into a complex organism. It is to see the vitality of cells and organisms in the physical and chemical dynamics of individual molecules and molecular motors. And because these dynamics are obviously physical and chemical, the mysteriousness of developmental processes seems to dissolve before one's eyes—even in the absence of a precise theoretical account of the kind a physicist might wish for. At the same time, because what one sees is not in fact the collective dynamics of an ensemble of molecules but rather the behavior of individual molecules and molecular assemblies in time, the molecules themselves come to take on a life of their own. This perhaps is what Marc Kirschner and his colleagues mean by molecular "vitalism."[51] It is also, I suggest, what lies behind an apparent resurgence of the language of natural history in the current literature—particularly in the frequency with which one finds recourse to analogies with ecology and the study of animal behavior.[52]

On another level, the kinds of observation that have now been made possible have undoubtedly played a major role in focusing the attention of biologists on the importance of coordinated cellular processes in development. "We have always underestimated cells," writes Bruce Alberts:

> Instead of a cell dominated by randomly colliding individual protein molecules, we now know that nearly every major process in a cell is carried out by assemblies of 10 or more protein molecules. And, as it carries out its biological functions, each of these protein assemblies interacts with several other large complexes of proteins. Indeed, the entire cell can be viewed as a factory that contains an elaborate network of interlocking assembly lines, each of which is composed of a set of large protein machines.[53]

Kirschner and colleagues argue for a similar shift in perspective, away from the structure and activity of genes to the importance of the activity of protein assemblies: "As it is now clear that gene products function in multiple pathways and the pathways themselves are interconnected in networks, it is obvious that there are many more possible outcomes than there are genes. The genotype, however deeply we analyze it, cannot be predictive of the actual phenotype, but can only provide knowledge of the universe of possible phenotypes."[54]

To be sure, technology for obtaining new visual access to in vivo development is only one of the many recent advances in molecular and cell biology that are responsible for this shift. Yet its role in bringing about an increasing awareness both of the complexity of cellular dynamics and of the importance of these dynamics in specifying the actual phenotype of the organism has surely been crucial. More specifically, for illustrating just how direct observations of living systems have contributed to a move away from a perspective of strict genetic determinism, I suggest that the recent investigations of Fraser and his colleagues of neural crest cell migration in intact embryos with confocal microscopy provide a particularly instructive example.

For some, the very diversity of mechanisms that are seen to come into play in determining cell fate serves as a warning against premature theorizing. Traditionally, theoretical models in biology have been associated with the singling out of one particular dynamic, of one particular kind of mechanism, and the message drawn from increasing visual access underscores, once again, how far the range of biological innovation exceeds the range of human imagination. Indeed, for many, merely being able to identify the mechanisms involved suffices as an explanation. For others, however, the identification of new mechanisms points in just the opposite direction, serving as a spur to new theoretical formulations. In particular, it has encouraged a conspicuous resurgence in efforts to construct new mathematical and mechanical models among those who want to know how these mechanisms actually work in physical terms. As Mehta and his colleagues write,

> A new era of biomechanical studies has been ushered in by the development of optical and mechanical probes that are sensitive enough to make measurements on single biological molecules . . . A general goal in molecular biophysics is to characterize mechanistically the behavior of single molecules. Whereas past experiments required model-dependent inferences from ensemble measurements, these new techniques allow a direct observation of the parameters that are relevant to answering the following questions: How does a protein move? How does it generate force? How does it respond to applied force? How does it unfold?[55]

These questions are of obvious interest to anyone who wishes to apply biological design principles to engineering, or to those who remain unsatisfied in the absence of a physical (mechanistic) account, but it remains far from obvious in what sense they are *biological* questions. And for many biologists they are not only beyond the range of their expertise but, for the most part, also beyond the range of their interests. Thus far, at least, analysis of molecular kinematics has failed to tell them what they most wish to know.

A third and final consideration in assessing the impact of new visualization technologies is their influence on readers and spectators outside the narrow corridors of the research laboratory. Four-dimensional representations have become a central component of presentations of new results to colleagues in seminars and conferences, and the accessibility of video clips on the Internet and the use of CD-ROMs as adjuncts to traditional journals have brought the observation of intra- and intercellular dynamics to yet larger audiences of specialists and nonspecialists.[56] Accordingly, it is not only the researcher who has the opportunity to watch these processes unfold in living time: the remote spectator, too, who is often far removed from the site of "direct" observation, has "virtually" the same opportunity.

Fifteen years ago, Steven Shapin and Simon Schaffer introduced the felicitous term "virtual witness" to describe the role of such remote spectators, and the expression has since become part of the basic vocabulary of historians of science. But the CD-ROM, I suggest, gives new meaning to the notion of "virtual witnessing."[57] For, as anyone who has seen high-quality video representations of biological development will recognize, the experience is at once thrilling and compelling, and in ways that traditional representations can scarcely begin to rival. So lifelike can the animated spectacle be made to appear that it induces a powerful sense of firsthand witnessing, the conviction that one is watching "life itself." Thus, the very technology that has so vastly increased our visual access to the inner workings of living organisms also has an ironic side-effect—namely, that our perception of "real time" comes to be more and more closely assimilated with our perception of "reel time." As Gregory Mitman argues, the pun serves to underscore the increasing elusiveness of the "real," even as our grasp of the objects to which that term refers becomes ever more firm.[58] Today, the term "visual reality" is used more or less indiscriminately to refer to the video-enhanced and computer-reconstructed images the microscopist sees, to the display of these images the remote spectator sees, and to visual representations of molecular processes that no one has ever been able to see.[59]

Such a conflation between the real and the virtual inevitably generates serious questions for philosophers, but such questions are surprisingly absent from the biological literature. Even more remarkably, concerns about artifacts of the kind that so openly haunted the earlier history of microscopy seem also to have been effectively eclipsed by the availability of images that look so dramatically and so compellingly alive.[60]

CHAPTER EIGHT

New Roles for Mathematical and Computational Modeling

> . . . when they come to model Heaven
> And calculate the stars, how they will wield
> The mighty frame; how build, unbuild, contrive
> To save appearances; how gird the sphere
> With centrick and eccentrick scribbled o'er,
> Cycle and epicycle, orb in orb.
>
> Milton, *Paradise Lost* (1667)

ver the last two decades, the work of historians and philosophers of science has undergone an unmistakable shift in focus. Today, the technical practices of a science, and not its theoretical achievements, command the lion's share of attention. Workers in the field now contend that what best distinguishes particular scientific disciplines from one another are the particular techniques and instruments employed, the "tool boxes" available.[1] But as I think back about my own socialization as a theoretical physicist in the late 1950s and 1960s, I realize that my induction into this professional culture preceded my learning any particular techniques. It began, rather, with learning an ethos. Long before I had any idea that theoretical physics was what I wanted to do—indeed, before I even knew who or what a theoretical physicist was—I was already being schooled in the founding cultural axioms of a discipline the name or even existence of which I had as yet no inkling. Above all, one might say that I was learning the right way to think about "thinking": how to recognize a certain kind of mental activity as its highest form, distinguished not only from feel-

ing and doing but also from other kinds of activities we usually think of as mental—for example, mere calculating, a form of work an early mathematics professor of mine quaintly referred to as "plumbing."

Furthermore, once I had developed a proper appreciation of the domain of "pure thought," inhabited by those who seek only knowledge, understanding, and truth, the demarcation of a parallel domain of "pure theory" seemed to follow as a matter of course. In this ideal universe, the term theoretical is cleansed not only of practical interests (under which the various forms of making and doing are all subsumed) but simultaneously of any implication of intent or directionality. This distinction, in short, is one upon which a discipline called theoretical physics depends, dividing those concerned with knowledge of deep truth (theorists) from those concerned with problems of making and doing (experimental or applied physicists).

Odd as it now seems to me, I did not regard this taxonomy of human activities as elitist. That I did not is itself an indication of the extent to which I was already into the process of cultural assimilation. Rather than elitist, I saw it as just a bit of elementary geography I needed to learn in order to embark on a quest that was itself so compelling as to effectively silence any such quibble. The quest was for a form of thought so pure and so powerful that it could forge an unmediated union between mind and nature and enable us, as Plato had written, to "grasp the essential nature of things." Like the promise of other unions, it drew me with the power of love. If such a quest required a distancing of the social and experiential world, as it surely did (I vividly remember the disorientation evoked by excursions into "ordinary life," the sharp sense of inhabiting a different world, and my puzzlement about how such a chasm had come to be), then so be it. It seemed a small price to pay for initiation into so rarified a culture, a culture in which I would be taught to carve the natural world at its true and proper joints, and to do so by wielding Plato's proverbial knife in such a way that it would keep its edge forever.[2]

It took me many years to appreciate the magical appeal of this vision as being just that, magical, and the vision itself as phantasmagoric. As recent studies of scientific practice have made clear, there is no such thing as pure thought, although the consequences of believing that there is are real enough.[3] But only after I began to think about the meaning of theory in contemporary biology did I come to see just how inadequate, and how misleading, such a notion of pure thought is as a description of the actual practices responsible for the growth of scientific understanding—in biology, in many other natural sciences, and even in theoretical physics itself.

Accordingly, as I approach the question of how new technologies have begun to change both the status and meaning of mathematical modeling in developmental biology, and hence the explanatory practices of that field, I want to speak not of theory but of theoretical work. Despite experimental biologists' traditional lack of interest in mathematical models and abstract theorizing, it would be a serious mistake to conclude that theory has had no place in the history of modern biology. In fact, certain kinds of theoretical work have always been in evidence and, in fact, essential to the practice of their science. A simple distinction between the two meanings of theory I have in mind is provided by a standard dictionary definition. Theory is defined both as "the analysis of a set of facts in their relation to one another" and as "the general or abstract principles of a body of fact."[4] Biologists' disavowal of theory is clearly directed at the second definition; by the first, theory is in fact constitutive of their very endeavor. Of necessity, both the interpretation of experimental data and the design of new experiments depend on extensive and sophisticated theoretical analysis of the possible relationships that can be brought into consistency (or inconsistency) with the data at hand.

That data do not speak for themselves is well known. But we have yet to develop an adequate appreciation of the kinds of work involved in giving voice to particular kinds of data. In molecular analyses of developmental genetics, for example, observed effects are given meaning through the construction of provisional (and often quite elaborate) models formulated to integrate the new data with previous observations from related experiments. As the observations become more complex, so too do the models that biologists must construct to make sense of their data. And as the models become more complex, the computer becomes an increasingly in dispensable partner in their representation, analysis, and interpretation.

Of the many differences made by this turn to computers in biology, one of the most prominent has been to render the conceptual work performed by these models easier to recognize as theoretical. Yet the new kinds of mathematical model biologists develop and explore are far more closely grounded in experimental realities and hence far more acceptable to working biologists than any of the mathematical models that had earlier been possible. And the extraordinary representational powers of computer graphics bring the results of these computations within easy reach of biologists with little or no mathematical expertise.

The computational models biologists are constructing may be qualitative or quantitative, and they may look quite different from the kinds of models traditionally employed in theoretical physics. Yet here too conventions are changing, and as a consequence of the growing reliance on similar kinds of models in physics, categorical distinctions between modeling practices in the physical and biological sciences are becoming ever more difficult to draw. As I hope to show in this chapter, a focus on the role of computational models in contemporary biology—asking what they are and what functions they are intended to serve—illuminates not only past differences that have divided the epistemological cultures of biology and theoretical physics but also some of the ways in which ongoing changes, driven in large part by technological developments, have begun to effect at least a semblance of cultural convergence between these disciplines.

Finally, and of particular relevance to the current concerns of many philosophers of science, the uses of these models in biology provide particularly clear examples of theoretical practices in which neither thought nor theory is ever pure and where the mix of theoretical and practical, conceptual and material, work is both conspicuous and inescapable. They provide especially good examples of recent arguments for regarding models as tools or instruments for conceptual development, and as such provide clear support for the challenges to traditional distinctions between theory and practice that philosophers like Margaret Morrison, Nancy Cartwright, and Mary Morgan have advanced.[5] Going one step further, I will also argue (both here and in the next chapter) that these models can at the same time serve as tools or instruments for material change, as guides for *doing* as much as for thinking. As such, they should be immune to the suspicion that the argument which has been put forth for models as mediators between theory and things relies too heavily on a "merely" metaphoric notion of tools and instruments.[6] Indeed, the very notion of merely metaphoric is put to the lie by the fact that metaphors, like models, can themselves function as tools for material intervention.

In an effort to ground these remarks and to better understand the meanings and functions of computational models in contemporary developmental biology, I choose two recent examples that have attracted particular attention among experimentalists in the field, and I examine each in some detail. The first is a model of (and arguably, also *for*) the regulation of gene activity—that is, of the rate at which a sequence of DNA that codes for a key protein is transcribed. This model was developed by Eric Davidson and his colleagues at the California Institute of Technology. The second, constructed by Garry Odell and his students at the University of Washington, is a model of networks of gene interactions that can give rise to developmentally robust patterns of segmentation in *Drosophila* and related insects.[7] In each case, I ask about the character of the model, the process by which it was elaborated, and its explanatory functions. The two examples differ in a number of respects—in their reliance on metaphor, in their use of computers, and in their relation to theoretical concepts, experimental work, and technological innovation. Even though they do not represent the full extent of computational practices in contemporary developmental biology, they illustrate both the range of meanings the term "computational model" has come to assume, and the variety of uses to which such models can be put.

A Model Of and For the Regulation of Genetic Transcription

Eric Davidson has been a leading figure in molecular developmental biology ever since the 1960s and has played a critical role in the growth of the field.[8] For the past decade, he and his colleagues have been studying the structural organization of the transcriptional promoter of *Endo16,* a developmental gene encoding a multi-functional protein of fundamental importance for the development of the sea urchin embryo. In 1998 their efforts culminated in a widely noted article in *Science* describing "a quantitative computational model . . . that reveals the logical interrelations hard-wired into the DNA" of this gene.[9] In an accompanying commentary entitled "Promoter Logic," Gregory Wray explains its special significance: "In spite of considerable investigation of the function of animal promoters, general principles have remained frustratingly elusive. There is little logic apparent in the organization of regulatory elements . . . It therefore comes as a surprise to discover a promoter that operates in a logical manner."[10]

In particular, the promoter of *Endo16* seems to be a "genetic computer" that "acts like a logic circuit (top) to determine expression of the gene (bottom)." Wray writes, "The 'program' that runs this tiny computer is directly encoded in DNA as regulatory elements; its inputs are single molecules whose composition varies in time and among various cells of the embryo, and its output is a precise level of transcription." Although he expresses doubt that the model will prove applicable to other promoters, he nonetheless concludes, "The results of Yuh and colleagues offer the hope that the seemingly haphazard operation of animal promoters might become more comprehensible to developmental and evolutionary biologists alike."[11]

What exactly is the meaning of "model" here? How does it relate to the modeling practices traditionally employed in the physical sciences? And how does this model contribute to our understanding of development? Figure 7 shows a cartoon version of the model; Figure 8 gives it as depicted in the original paper. In neither representation does one see any of the differential equations that form the heart of the quantitative models one typically encounters in physics or in most of the earlier efforts in mathematical biology. What Figure 7 provides is a schematic sketch of a logic circuit, and Figure 8 provides the specifics for a program as it would be written for simulation on an actual computer.[12] This is not a program written to simulate the behavior of a model that has been elsewhere specified (such as a set of otherwise intractable differential equations such as the Navier-Stokes equations): the logic circuit or program *is* itself the model. And indeed, in conversation with several physicists, I was informed that what we have here is not a "model" at all but "merely a description" or "just an algorithm"—nothing more than a "schematic representation" of a set of experimental results that would otherwise be too unwieldy to manage. I argue, however, that such economy of representation is a primary function of all models. Moreover, this model, despite its very different appearance, serves most of the conceptual functions of conventional theoretical models in physics, as well as certain distinctly practical functions: it suggests experiments, it enables one to predict the consequences of particular interventions, and it prompts the posing of new questions.

The main sticking point for those whose notion of model is primarily informed by traditional examples in twentieth-century theoretical physics seems to be two-fold. The first has to do with generalizability: If the model is not generalizable—that is, is not applicable to other developmental genes—exactly what, and how, does it help us understand? The second, with its representational status: Are we supposed to take the notion of an analog device or computer built into the DNA literally? Clearly not. How then are we to take it? I will tackle these questions in turn, but first, we need to know a bit more about the model itself—what it is intended to explain and how it has functioned.

The biological issue is clear enough. Normal development depends on the activation of particular genes at the right time and in the right place. How is this achieved? More specifically, how is the activation or deactivation of *Endo16* (that is, the start and stop of its transcription) determined? Prior to the formulation of the model, experimental analysis of the regulation of this gene had revealed a wealth of information which, by itself, offered no obvious answer to the question at hand. In particular, it was known that:

(1) An approximately 2,300 base-pair sequence of DNA upstream from the site at which transcription begins suffices to regulate normal expression of *Endo16*. This is known as the promoter region.[13]

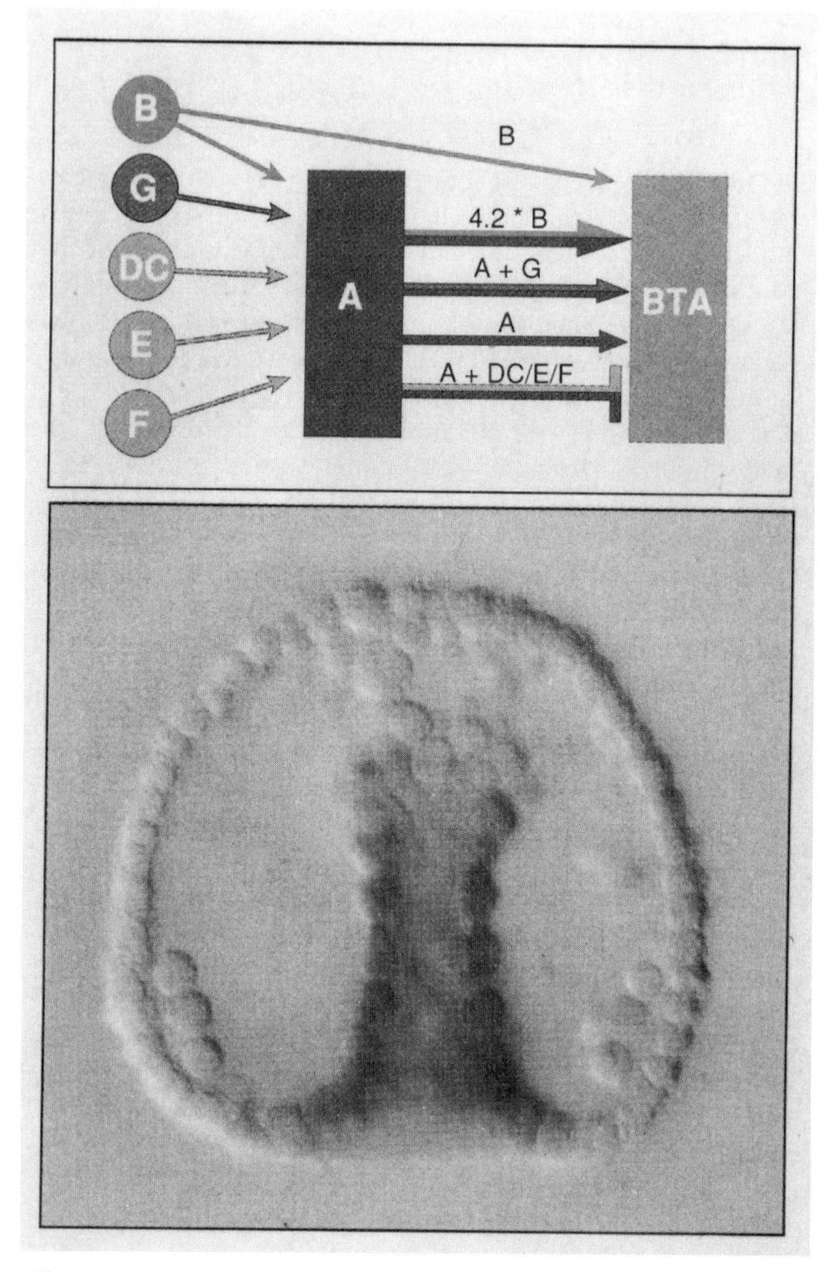

Figure 7. A genetic computer. The promoter of *Endo16* acts like a logic circuit (top) to determine expression of the gene (bottom). (Reprinted with permission from Wray, "Promoter Logic," copyright 1998, American Association for the Advancement of Science.)

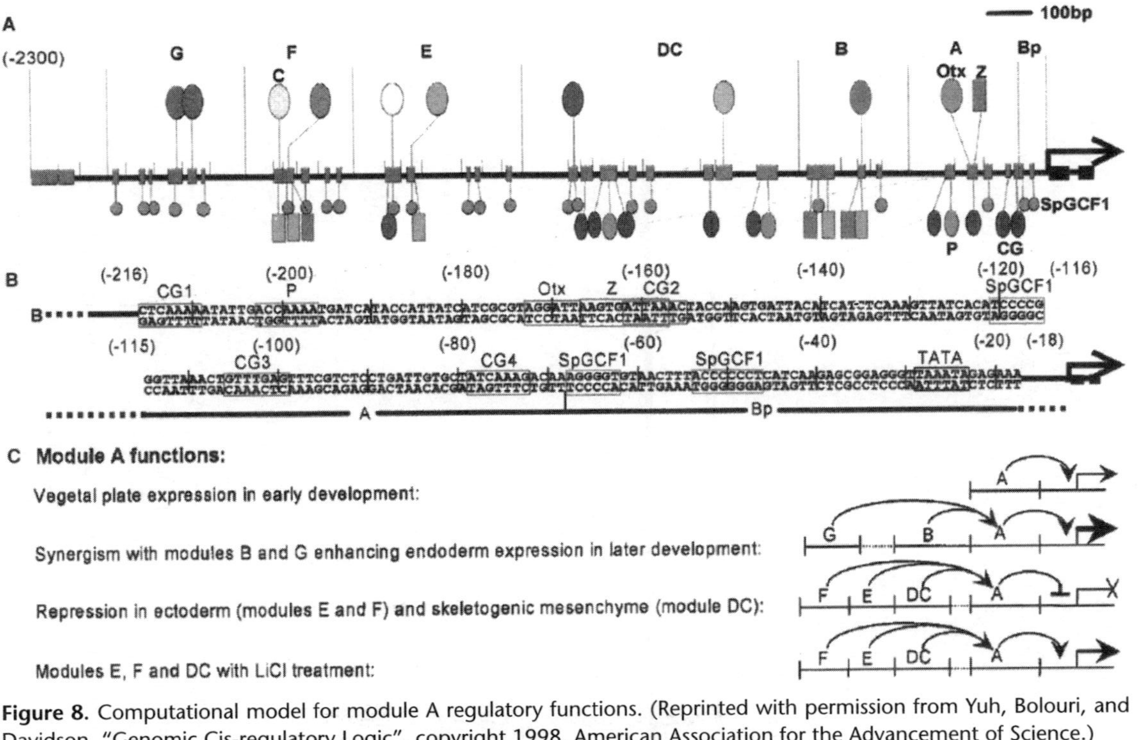

Figure 8. Computational model for module A regulatory functions. (Reprinted with permission from Yuh, Bolouri, and Davidson, "Genomic Cis-regulatory Logic", copyright 1998, American Association for the Advancement of Science.)

(2) Embryonic expression of *Endo16* is controlled by the binding of at least 13 different transcription factors (proteins distinguished by size and target site sequence but as yet unidentified) at over 30 highly specific binding sites along this region of DNA. In addition, over 20 binding sites had been identified for an additional transcription factor that is well known from other regulatory studies.[14]

The difficulty is not simply that this information is not sufficient but that it is, in another sense, too much. The identification of 50 binding sites implies the existence of a minimum of 2^{50} possible combinations (assuming only two states for each site).[15] While such a large number enables this gene to be acutely sensitive to its immediate environment, it creates an obvious problem for us in attempting to answer the question of how all this information is organized. What, in fact, would count as a satisfying explanation for how this information is translated into starting (or stopping) transcription? One possibility, of course, would be to simply itemize all 2^{50} combinations and experimentally determine which combinations are correlated with transcription and which are not, and then to search for the actual biochemical processes effecting such correlation. But the sheer magnitude of the number of sites makes this, a priori, an unsatisfying account.[16] Before claiming to "understand," we would need to organize this information into a pattern with a logic that is recognizable to us. This, I believe, is what Wray means when he writes of "discovering" that this promoter behaves "in a logical manner." And here, I argue, is where the formulation of a model of some kind becomes necessary. And this is precisely the point in their investigations at which Davidson and his colleagues began to articulate a rudimentary model.

Their first step rested on an early observation that the binding sites appear in clusters, and it was a crucially simplifying step: on the basis of this observation, they sought to correlate different parts (modules) of the promoter region with specific regulatory functions corresponding to particular stages of development.[17] Figure 9A shows the structural breakdown of the region, and Figure 9B shows a schematic representation of the correlation between structure and function. Success in demonstrating this correlation meant that they could drastically reduce the number of variables to the gross behavior of six modules (B–G) and focus their attention on the organization of the information from these modules in module A, the module most proximate to the starting point for transcription.

Yet even from this simplified reconstruction to the computational model shown in Figure 8 is hardly a small step. Figure 8 depicts the organization of only a single module (module A), in which the output of all other modules is integrated and transformed into a signal to the adjacent gene to start (or stop) transcription. The authors describe this module as "a central processor" or "switching unit," and its structure is depicted in a mode fully consonant with the guiding image of a computer. Development of the model required a tremendous amount of work—work that involved ever more fine-scale analysis of the function of individual binding sites, with a constant back and forth between the formulation of the model and the design and execution of experiments.[18] Indeed, throughout this process, the word "model" is probably best understood as a verb, with the authors as subject, and the experiments and the conceptual schematic as a single, unparseable, composite object.[19]

Only at the end of the process do we have a separable entity—a model as a noun, an entity that can be employed as a quasi-independent tool for designing new kinds of experiments, for posing new kinds of questions, and for guiding new kinds of manipulation of the system itself. For example, once various logical

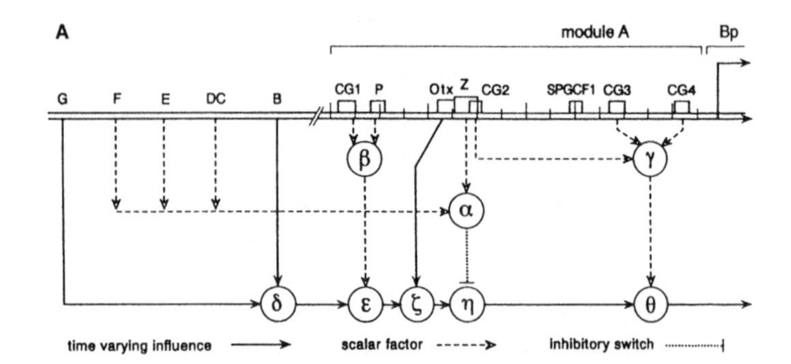

B

if (F = 1 or E = 1 or CD = 1) and (Z = 1) $\alpha = 1$	Repression functions of modules F, E, and DC mediated by Z site
else $\alpha = 0$	
if (P = 1 and $CG_1 = 1$**)** $\beta = 2$	Both P and CG_1 needed for synergistic link with module B
else $\beta = 0$	
if ($CG_2 = 1$ **and** $CG_3 = 1$ **and** $CG_4 = 1$**)** $\gamma = 2$	Final step up of system output
else $\gamma = 1$	
$\delta(t) = B(t) + G(t)$	Positive input from modules B and G
$\varepsilon(t) = \beta^* \delta(t)$	Synergistic amplification of module B output by CG_1-P subsystem
if $(\varepsilon(t) = 0)$ $\xi(t) = Otx(t)$	Switch determining whether Otx site in module A, or upstream modules (i.e., mainly module B), will control level of activity
else $\xi(t) = \varepsilon(t)$	
if $(\alpha = 1)$ $\eta(t) = 0$	Repression function inoperative in endoderm but blocks activity elsewhere
else $\eta(t) = \xi(t)$	
$\Theta(t) = \gamma^* \eta(t)$	Final output communicated to BTA

Figure 9. *Endo16* cis-regulatory system and interactive roles of module A. (A) Diversity of protein binding sites and organization into modular subregions. Specific DNA binding sites are indicated as blocks; modular subregions are denoted by letters G to A (Bp, basal promoter). (B) Integrative and interactive functions of module A. Module A communicates the output of all upstream modules to the basal transcription apparatus. It also initiates endoderm expression, increases the output of modules B and G, and is required for functions of the upstream modules F, E, and DC. These functions are repression of expression in nonendodermal domains and enhancement of expression in response to LiCl. (Excerpted with permission from Yuh, Bolouri, and Davidson, "Genomic Cis-regulatory Logic," copyright 1998, American Association for the Advancement of Science.)

functions (such as linear amplification and synergism) have been identified, a clear mandate is presented to the biochemist to find the actual proteins that can perform such functions. Similarly, the question of how such a complex mechanism might evolve poses a pressing theoretical problem for the evolutionary biologist, for it is difficult to see how the traditional answer of random mutation followed by selection could suffice.[20] Finally, the circuit diagram in Figure 8 provides a clear guide for altering the program, that is, it tells us how to modify the promoter region to change the pattern of gene expression.

However, we are still left with the question of generalizability. What can this model tell us about the regulation of other developmental genes, and hence about the logic of development more generally? In one sense, the answer is very little: this is a model developed for the regulation of a particular developmental gene, with little expectation that it will apply to many (if any) others, or even that it will offer any usable shortcuts in their explication. Such an answer will almost surely be disappointing to most physicists, but probably not to most biologists. In fact, what it tells the biologist may be more important than a general rule or principle—what it does is to exhibit a new dimension of variability in the structure of regulatory systems: as Davidson puts it, "It makes one realize how different they can be."[21]

Equally interesting is the question of the representational status of this model: How literally are we to take the metaphor of a "genetic computer"? Computer metaphors have been commonplace in biology for almost half a century. But until recently, they have been just that, metaphors which function without the expectation that they will approach literal correspondence. Computers may have been seen as functional analogs of biological systems but not as structural analogs: no one would have taken them as even approximately corresponding to the physical structures of DNA, cells, or organisms.[22] That is, not until the last few years, when a number of computer scientists began to look to molecular biology for new horizons in computer technology.[23]

Modeling a Robust Developmental Module

Garrett Odell is also an experimental developmental biologist, but, unlike Davidson, he had begun his career as a mathematical biologist. By the mid-1980s, however, despite a growing eminence in the field, he had lost confidence in the value of the contributions that a mathematician, not immersed in the complexities of real organisms, could make to our understanding of biological phenomena.[24] His loss of confidence was shared by a number of other mathematical biologists of his generation (some of whom were already forming research collaborations with experimental biologists), but Odell made a bolder move. In 1986 he gave up a tenured professorship in the Mathematics Department at Rensselaer Polytechnic Institute and moved to the Department of Zoology at the University of Washington. Here, in collaboration with Victoria Foe and others, he embarked on a series of experimental studies of early development in *Drosophila* and related insects, bringing to bear his long experience in theoretical and applied mathematics. He also attracted a group of graduate students equipped to pursue a new kind of mathematical biology, one in which experiment leads and the role of mathematics is relegated to servitude.

A problem of particular concern for Odell's group is segmentation in insect bodies, an interest growing out of a long-standing preoccupation with the question of how networks of gene interactions give rise to patterns of gene expression.[25] Although the precise mechanism by which segments are specified may vary radically across different insect orders, all insects possess homologous segments. Furthermore, recent studies show extensive conservation of segment polarity genes over the course of insect evolution of the segment, but not of other genes involved in earlier stages of the process (such as *gap* and *pair-rule* genes). This suggests that the segment polarity gene network is in some sense modular, for similar patterns of segment polarity gene expression appear to result from very different inputs. Indeed, the current presumption among researchers is that evolution works by rearranging such distinct and more or less autonomous developmental modules, each of which is associated with its own characteristic behavioral response to a range of different inputs. If so, what might such an evolutionarily stable module, one that is relatively insensitive to the details of input stimuli and at the same time resistant to known interspecific variations in development, actually consist of?

The paradigm case for insect development is of course *Drosophila*. Thanks in good part to the work of Christiane Nüsslein-Volhard and her colleagues, a remarkably clear picture had emerged by the late 1980s of the cascades of genetic regulation in *Drosophila* that are set in motion, prior to cellularization, by the differential distribution of maternal "morphogens." By the mid-1990s, the work of many other investigators resulted in an equally detailed description of events following cellularization.[26] After cell membranes are laid down, the spatial pattern of transcription products that had been established before cellularization serves as the initial input to a segment polarity network of genes, giving rise to a stable pattern of gene expression reiterated across 14 distinct (multi-cellular) domains, each corresponding to a single segment. The fact that, even among insects, the long syncytial stage of *Drosophila* development is anomalous makes it difficult to draw any general conclusions from the specifics of the early phases of this particular insect's development. But the events occurring after cellularization, especially those involving the segment polarity network, seem to be more general, and the possibility suggests itself that this network of genes may indeed constitute a developmental module of just the kind imagined.

To explore this possibility, and to determine "whether the known interactions among segment polarity genes suffice to confer the properties expected of a developmental module," George von Dassow, Eli Meir, Edwin Munro, and Garrett Odell set out to model the temporal dynamics of the network of genes and gene products already known to be involved, relying on computer simulations of their model to investigate its properties.[27] Their approach is a traditional one: rather than relying on metaphoric assimilation between organisms and computers, they see the computer as nothing more than a computational machine, and they describe the organism solely in terms of the genetic and biochemical properties that biologists have thus far elucidated. Furthermore, these properties are represented in the conventional form of continuous, time-dependent, differential equations that are only subsequently (and approximately) recast in terms of the discrete variables that computers can handle.

For the sake of computational tractability, the authors limited themselves to the five segment polarity genes that are believed to be most critical *(wingless, hedgehog, engrailed, patched,* and *cubitus interruptus).* Yet, in order to include the most firmly established interactions between and among even this reduced number of genes and their products, even when these interactions have themselves been highly simplified, a formidable number of coupled differential equations was still required (136, to be exact). Moreover, the equations include a large number of unspecified parameters (approximately 50) to represent the relevant half-lives, diffusion constants, binding rates, and cooperativity coefficients. In the absence of virtually any available information that could be used to fix the value of these parameters, they were obliged to make use of estimates based on random sampling over the range of biologically plausible values in order to obtain solutions of their equations.

But while their model is traditional in its use of differential equations to represent the system, it is still a far cry from the simplicity and elegance toward which models in the physical sciences have aimed in the past. Indeed, such a model would be unthinkable without the phenomenal number-crunching power of modern computers. It is cumbersome, messy, and, by itself effectively opaque to any kind of intuition (mathematical or otherwise).[28] Just as with so many of the models now employed for the analysis of such complex physical systems as fluids and weather conditions, interpretation and understanding depend crucially on the actual solutions generated by the computer. To be sure, Turing's early model of embryogenesis (discussed in Chapter 3) also relied on computation, at least for its full analysis, but in that case the most fundamental prediction of the model (namely, the onset of instability) could be directly obtained from the simplest sort of mathematical analysis. Not so in this case. In fact, the question posed here is a different kind altogether. These investigators ask not whether a developmental module with the expected properties can be imagined but whether the known interactions among segment polarity genes can, even in principle, suffice for such a module. As the authors put it, "Is there any set of parameter values for which the network model exhibits the desired behaviour, given realistic initial conditions?"[29]

The surprise, and possibly the principal value, of their model was that they were unable to find any such set. Despite extensive efforts, none of the parameter values they sampled produced patterns with the requisite stability. They concluded, therefore, that their initial set of interactions must be incomplete—that essential linkages were missing. The network as originally described was not, by itself, able "to explain even the most basic behaviour of the segment polarity network." In an attempt to remedy the problem, they added two additional interactions (both of which could find some grounding in experimental evidence) and found that, with these additions, the behavior of the system changed dramatically. Now, a parameter value chosen at random was observed to have a 90 percent likelihood of being compatible with the desired behavior, with the net result that a significant proportion of parameter sets ($0.9^{48} \approx 1/200$) was capable of generating stable patterns of gene expression that resemble those actually observed in the laboratory. Because a solution could be generated for almost any value of an individual parameter, and because they found no clustering among successful values in narrow subranges, they conclude that "the network's ability to pass our test is intrinsic to its topology rather than to a specific quantitative tuning. There are so many diverse solutions that the notion of a globally optimal parameter set makes no sense." Furthermore, they continue, "not only does the network topology embody many different solutions, but most solutions are highly robust to variation in individual parameter values." In other words, "the simplest model that works at all emerged complete with unexpected robustness to variation in parameters and initial conditions."

What can be learned from such efforts? One lesson is immediately obvious, and it is concisely stated by the authors themselves: "Biologists' maps of gene networks are rapidly outgrowing our ability to comprehend genetic mechanisms using human intuition alone, as shown by our initial failure. Our results reveal holes in the current understanding of segmentation."[30] Moreover, their model provides a guide for filling these holes. And from their success, Peter Dearden and Michael Akam draw a sweeping conclusion. As they write in their accompanying ("News and Views") article, "A new mathematical biology is emerging . . . Building on experimental data from developing organisms, it uses the power of computational methods to explore the properties of real gene networks." Of particular importance to Dearden and Akam is the fact that, unlike earlier efforts in mathematical biology, here we have a model "so closely tied to real data that its inadequacies immediately define a programme of experimental work to test those assumptions." Furthermore, the very fact that the results of the model computations are counter-intuitive makes it "certain that modelling will be essential to make sense of the flood of new data. But," they continue, "this will not be elegant theoretical modelling: rather, it will be rooted in the arbitrary complexity of evolved organisms. The task will require a breed of biologist-mathematician as familiar with handling differential equations as with the limitations of messy experimental data. There will be plenty of vacancies, and, on present showing, not many qualified applicants."[31] Those now being trained by Odell—as von Dassow, Meir, and Munro were—are still the exceptions, but if Dearden and Akem are right, they will not be so for long.

But the interdisciplinary negotiations required remain tricky. What may well be the most interesting finding from this exercise in a new mathematical biology is scarcely mentioned by Dearden and Akam, and it goes considerably beyond the tweaking of our current understanding of a particular biological process like segmentation. In their discovery of the surprising robustness of any network capable of generating the basic pattern observed in this process, the authors seem to have stumbled on a phenomenon of vast biological importance, and it is one that has been as invisible to direct experimental inquiry as to the simplified mathematical models constructed by earlier generations. "Why," they ask, "should the segment polarity mechanism be so robust?"[32]

Elsewhere, I have argued that genetic analyses of development have long ignored the importance of robustness in developmental processes.[33] But now, this phenomenon emerges from the analysis of genetic networks as it were, spontaneously, and the authors have been led to ask some crucial questions: Why do embryos require so much buffering? What kinds of evolutionary pressures might have given rise to such an unexpected property? Indeed, consideration of these questions, based on the lessons of their modeling efforts, now leads them to conclude that "robust gene networks are the only networks natural selection can evolve."[34] Perhaps their models have led them to the identification of a fundamental feature of biological reality, and not just the working out of hypothetical principles.

The Healing of a Classical Divide

On April 7, 1998, the Institute for Advanced Study in Princeton announced the formation of a new program in theoretical biology. As the founders explained in a press release:

> The use of mathematical ideas, models, and techniques in the biosciences is a rapidly growing and increasingly important field. Applied mathematicians have traditionally used mathematical methods to address a wide range of problems in the physical sciences, especially physics and engineering, in the belief that the underlying laws of physics are of a precise nature and therefore capable of being described mathematically. While the physical sciences have had this mathematical/theoretical tradition from their beginnings, biology has had a different history . . . [It] has been more focussed on laboratory work. However, several areas of biology have gradually developed an understanding of the important role that mathematical approaches can play. Such approaches are often in the hands of people who collaborate with experimentalists, but do not themselves work in the laboratory.

Such indications of a historical shift in the status of mathematical/theoretical modeling in the biological sciences are not hard to find, and they come from a variety of different sources. What is more difficult to find in this particular statement is any sign of the depth of the changes that seem to be taking place. In many ways, it reads as though it was written in an earlier era, by authors still firmly steeped in the epistemological culture of theoretical physics.

That cultural location is revealed in a number of premises the statement takes as self-evident—most conspicuously, in its equation (or conflation) between mathematical and theoretical; in its acceptance of a well-demarcated divide between theory and experiment along with an institutional affirmation of this divide ("participants in this program will not themselves work in the laboratory"); and, finally, in its casting of biology as an underdeveloped culture which has only "gradually developed an understanding of the important role that mathematical approaches can play." Not only would these assumptions be far from obvious to working biologists but also they would seem both wrong-headed and belittling. Here, in the confident and uncritical assertion of the program's founders, biologists would find the unmistakable voice of a theoretical physicist, speaking out of his or her own particular (and largely twentieth-century) history, and seeking to impose the cultural givens of that history on their own discipline, both immediately and clearly recognizable.

Similar strains can be heard in another programmatic statement that was issued by a National Science Foundation–sponsored workshop that Simon Levin, a mathematical biologist, convened in 1992 "to explore current challenges and opportunities at the interface of Mathematics and Biology."[35] Although the report describes a number of judicious efforts on the part of the committee to chart a new path for the field, one that would avoid repeating past mistakes, it too begins with the canonical analogy between the dawn of modern physics in the early seventeenth century and the state of biology at the end of the twentieth century: "This is the stage in which biology finds itself today, poised for the phase transition that comes with the total integration of mathematical and empirical approaches to a subject. Many branches of biology are virtually devoid of mathematical theory, and some must remain so for years to come. In these, anecdotal information accumulates, awaiting the integration and insights that come from mathematical abstraction."

Very quickly, however, the authors distance themselves from claims that predecessors like Rashevsky had made by specifying at least one important difference and emphasizing the need for experimentalists and theorists to work together: "In other areas, theoretical developments have run far ahead of the capability of empiricists to test ideas, spinning beautiful mathematical webs that capture few biological truths. This report eschews such areas, and instead focuses on those where the separate threads are being woven together to create brilliant tapestries that enrich both biology and mathematics." Also, the report reminds us of the unique opportunities that "have surfaced within the last ten to twenty years" as a consequence "of the explosion of biological data with the advent of new technologies and because of the availability of advanced and powerful computers that can organize the plethora of data."[36] Such new opportunities, the authors argue, demand new research strategies, emphasizing interdisciplinary collaboration between biologists and mathematicians that aim for a pooling of expertise rather than for a "physics of biology."

Above all, they conclude that such interdisciplinary efforts will require new structures of funding:

> Flexibility by the funding agencies to the needs at this interface is essential. Cross-disciplinary teams of researchers should be encouraged and appropriate methods for review of proposals should be developed . . . Because the subject lies between traditional disciplinary areas, its support often "falls between the slats" at funding agencies. We urge that specific mechanisms be developed to recognize the unique character of the subject and to provide the support that will

> foster the development of work that truly can make contributions both to biology and to mathematics.

As it happens, their urgings dovetailed nicely with other needs that were then coming to be felt. The first and most immediately pressing such need grew out of the Human Genome Initiative, and the relevant funding agencies (especially the National Institutes of Health and Department of Energy) responded quickly to the demands of genomics research by spawning a number of new initiatives in bioinformatics and computational biology. But the pressure for a more general response to the need for interdisciplinary collaborations in contemporary cell and molecular biology continued. In the spring of 1996, a second workshop was convened at the NSF to call attention to "the tremendous potential of mathematical and computational approaches in leading to fundamental insights and important practical benefits in research on biological systems" and to "raise the level of awareness at the National science Foundation and other funding agencies on nurturing computational and mathematical research in the biological sciences."[37] This time around, the response has been unambiguous. Over the last few years, the NSF has launched a number of large-scale initiatives aimed specifically at realizing this potential,[38] and Mary Clutter estimates the proportion of funding for mathematical and computational research coming from the Biological Division in the year 2000 at approximately 13 percent.[39] Indeed, the complaint most frequently heard today concerns a paucity of qualified applicants, rather than a paucity of funding.[40]

In turn, the increase in resources is reflected in a correspondingly exponential growth in the rate of publication in the areas of mathematical and computational biology (with many of these articles appearing in such journals as *Science, Nature,* and *Proceed-*

ings of the National Science Foundation—journals that are routinely read by experimental biologists), and in the formation of new programs in mathematical or computational biology at many if not most major universities.[41] Of particular significance is the fact that, unlike their precursors, a number of these programs are housed in departments of experimental biology rather than in mathematics or computer science departments.

In these new settings, mathematicians and computer scientists do not simply collaborate with experimental biologists, they become practicing biologists. Conversely, and thanks in part to the rise in computer literacy and in part to the development of user-friendly computer programs that bring the techniques of mathematical analysis within the grasp of those with little or no conventional training in the subject, biologists need no longer simply hand over their questions and data to others; now, they can build (either by themselves or as active participants) their own mathematical/theoretical models.[42] The net effect is the beginning of a new culture in biology, at once theoretical and experimental, and growing directly out of the efforts of workers whom Dearden and Akam describe as "a breed of biologist-mathematician as familiar with handling differential equations as with the limitations of messy experimental data."[43]

If, as Dearden and Akam claim, "a new mathematical biology is emerging," it is one that bears only passing resemblance to the discipline pioneers like Thompson, Rashevsky, or Turing had sought to found. It promises not so much a vindication of past failures as a transformation of the methods, the aims, and the epistemological grounding of past efforts. The premises of the new mathematical biology depart not only from the traditional assumptions still so clearly evident in the IAS statement but even from the modified (and transitional) formulations of Levin's 1992 committee. Most conspicuously, they diverge on the meaning of "mathematical" and its relation to "theoretical," on the relation between theory and experiment, and in the respect (or lack thereof) accorded to the achievements of past and current biological research.[44] Of course, now as earlier, the term a mathematical biology refers to an extremely heterogeneous array of efforts, encompassing a wide domain of uses, applications, and kinds of mathematics employed. But if we restrict ourselves to Dearden and Akam's use of that designation and confine ourselves to the question of the role that mathematical approaches can play in contemporary developmental biology, we find here in particular the new conjunctions, in relation to every one of these issues, that demand the reformulation of traditional categories.

Without question, part of the impetus for new conjunctions in biology, just as in other scientific disciplines, comes from the increasing reliance on computational models; and a substantial philosophical literature on the conceptual impact of these new practices has already begun to accumulate. Peter Galison, for example, has described the use of computer simulations in the physical sciences as a practice that stands "in a novel epistemic position within the gathering of knowledge"; intermediate between experiment and theory, it appears to be fundamentally incommensurate with earlier categories.[45] And to some extent, the examples discussed above bear his conclusion out. While I have argued elsewhere that the uses of computers in scientific modeling are too varied to warrant easy assimilation, taken together, they have clearly led to new meanings of both mathematical and theoretical.[46] In fact, use of the term computational model as a virtual synonym for mathematical model has become commonplace, whether or not traditional methods of mathematical representation or analysis are involved. In many cases, computational analyses altogether displace the need for mathematics (at least in the strict sense of that term), while nonetheless still qualifying as both mathematical and theoretical.[47] More specifically, the convergence of theory and experiment that Galison observes in computer simulations is well exemplified by the studies of "artificial life" I turn to in Chapter 9. But it is a radically different kind of convergence from that displayed in the new mathematical biology discussed in this chapter. Here, the meaning of theory has shifted, and obviously so, but the term experiment remains unambiguously restricted to the domain of physical, material manipulations.

Crucial to this difference is the fact that the technological developments driving the convergence of theory and experiment in this context come not only from computers but also, and at least equally, from new tools for manipulating real biological materials and from the sheer quantity of data which these tools have generated. Sylvia Spengler's observation that "the biology community requires extensive, integrated computational facilities to handle the wealth of data generated" is echoed by many others, just as is the observation by von Dassow et al. that "biologists' maps of gene networks are rapidly outgrowing our ability to comprehend genetic mechanisms using human intuition alone."[48] Sydney Brenner writes, "There seems to be no limit to the amount of information that we can accumulate, and today, at the end of the millennium, we face the question of what is to be done with all this information." Furthermore, he is convinced "that the prime intellectual task of the future lies in constructing an appropriate theoretical framework for biology." But what is meant by a theoretical framework? Brenner acknowledges that "theoretical biology has had a bad name because of its past." But even so, and "even though alternatives have been suggested, such as computational biology, biological systems theory and integrative biology," he has "decided to forget and forgive the past and call it theoretical biology."[49] And why not?

The crucial point is that the phenomenal advances in molecular biology, particularly since the advent of recombinant DNA, have generated the need for new kinds of analysis. That computers have come to play so prominent a role in meeting this need is equally a consequence of history—in particular, of the rapid developments in that area. The result is a meshing of needs and resources in which we can begin to recognize the marks of a new epistemological culture. The particular examples I discuss in this chapter were chosen largely because of the interest they have generated among biologists not normally associated with the practices of mathematical or computational biology. Yet even by that criterion, they are far from exceptional, for models of similar kinds are proliferating in the current literature.[50] It hardly matters whether they go under the name of theoretical, computational, or mathematical biology, for the fact is that these very efforts are giving rise to new meanings for all of the relevant words.

First, the theoretical biology for which Brenner calls grows directly out of and remains inextricable from the experimental biology that has summoned it into being. Second, the traditional links binding "theoretical" to "mathematical" have been substantially weakened by the powers of computational analysis. And third, biology itself no longer stands in any kind of opposition to physical science. Brenner suggests that what still distinguishes biological systems from conventional physical-chemical systems is that, "in addition to flows of matter and energy, there is also flow of information. Biological systems are information-processing systems and this must be an essential part of any theory we may construct."[51] Today, however, physical systems too are coming to be thought of as information-processing systems. Thus, the very novelty that Brenner claims for the subject of biology can now work to bring his science into even closer proximity with the physical sciences.

Theory and Practice in Contemporary Molecular Biology

A close admixture of conceptual and material tools has characterized experimental biology throughout the past century. What lends the mix in contemporary molecular biology its principal novelty, I argue, is the arrival of new technologies. The computer vastly extends the range of tools available, while the techniques of recombinant DNA make such an extension virtually mandatory. But there is more, for the recombinant DNA revolution has also added new horizons to the very meaning of words like tool and practice. In the past, the possibilities for intervening in the course of biological development—without causing its disruption—were limited to control of the mating process and, only under special circumstances and in relatively isolated cases, to the manipulation (cutting and pasting) of body parts. Now, however, and for the first time in biological history, it has become possible to directly intervene in the internal dynamics of development without interrupting that process, and to do so on the molecular scale.

Techniques of recombinant DNA provide us with the means to target and alter specific sequences of nucleotides in ways that turn genetic markers into handles for effecting specific kinds of change. They have made genetic engineering a reality and, hence, a business. And in doing so, they have inevitably led to changes not only in the practices of biology, but also in the "practicality" of that science. The advent of recombinant DNA technology has brought contemporary molecular biology into striking accord with the notion of a "practical science" that the philosopher R. C. Collingwood, borrowing the term from Aristotle, put forward more than half a century ago.[52]

Collingwood began by drawing a sharp distinction between theoretical sciences (like physics) and practical sciences (taking the medical sciences as an example) and argued that explanation takes on an entirely different character in these two different kinds of science. Above all, he wrote, they differed in the meaning given to causal propositions. In the practical sciences, "The causal propositions which it establishes are not propositions which may or may not be found applicable in practice, but whose truth is independent of such applicability; they are propositions whose applicability is their meaning." For events in which a practical interest is taken, "the 'condition' which I call the cause . . . is the condition I am able to produce or prevent at will."[53]

There are good reasons for accepting his description as appropriate for much of contemporary molecular biology[54]—and perhaps even for much of the history of physics.[55] But in contradistinction to Collingwood, I take the distinction between theoreti-cal and practical to be far from obvious, and especially so when we consider the wide ranges of meaning of "practice" and "practical." Thus far, I have focused almost entirely on intersections between theoretical and experimental work, both of which activities depend on a variety of practices that must be "practical" if they are to be of use to the scientists who deploy them. But part of their success also depends on the practices of others engaged in allied efforts (both scientific and engineering) and on the ability of these others to find the work useful to their own needs and in their own contexts. In short, just as the distinction between theoretical and experimental begins to dissolve upon examination, so too does the distinction between pure and applied. The applications to which a theoretical description may lend itself can cover a large ground, often far removed from the intentions of the original investigators. They may show up in other labs, in other departments, or in ventures outside the academy altogether—in purely commercial settings, for example. When these other domains of applicability are taken into account, the word "practical" clearly takes on a far more general meaning than Collingwood had in mind.

Admittedly, neither Davidson nor Odell appear to have an interest in the use of their models for anything other than a better understanding of developmental dynamics. Nevertheless, other uses are not hard to imagine. For example, the model constructed by Davidson and his colleagues for the structure of the promoter for *Endo16* provides a clear map for intervening in that structure, thereby enabling specific alterations in the regulation of *Endo16*. In fact, the ability to do so was essential to the development and refinement of their model. Once the model has been refined, however, it can equally well serve as a guide to implementing modifications in the regulatory circuitry toward other kinds of practical ends. There may be little call for the commercial production of the protein (or proteins) encoded by *Endo16*, but it is not hard to find other proteins, under equally complex regulation, for which a great demand does exist. Models of a similar kind, constructed for the regulation of these other proteins, would clearly have enormous practical value in today's increasingly biotech economy.[56]

The metaphor of a genetic computer can also be argued to have practical ramifications, especially as employed in the emerging domain of biological computing. But I defer discussion of the engineering uses of new models and metaphors in developmental biology to the next chapter, where I focus more directly on the implications of new conjunctions between biology and computers for computer scientists.

CHAPTER NINE

Synthetic Biology Redux—Computer Simulation and Artificial Life

simŭlo; v. I. *To make* a thing *like* another; *to imitate, copy* . . . II. *To represent* a thing as being which has no existence, *to feign* a thing to be what it is not.

Lewis and Short, *A Latin Dictionary* (1958)

lmost a century ago, Stéphane Leduc sought to explain the origin and development of life through the construction of artificial organisms that looked like, and even seemed to behave like, real organisms. He induced the formation of these lifelike figures with ingenious uses of India ink and chemical precipitates, and he dubbed his enterprise "synthetic biology." Both Leduc and his readers clearly understood that term's double meaning.[1] Here was a biology that was synthetic in the sense both of being artificial and of providing at least a stepping stone toward what his colleague A. L. Herrera described as its long-range objective, namely, the synthesis of living matter. We would probably use the word "simulation" to describe such efforts, relying on the double meaning this word now (once again) connotes, but in Leduc's time, it was its negative and clearly pejorative sense that had come to predominate. Tied so closely to deception, falsity, pretense, the meaning of simulation conflicted too blatantly with the aims of science to qualify for Leduc's purposes. In fact, it was only in the context of war-related research in the 1940s that simulation began to recover its original (and potentially positive) sense of imitation that would bring it into alignment with Leduc's use of synthetic.

The first modern uses of simulation as a productive term (productive, that is, by virtue of its semblance to the real) are to be found in descriptions of pilot training programs in which electrical and electronic analog devices were employed to mimic the behavior of real-world phenomena.[2] Here, the term comes to be used as a technique for gaining experience. At the same time, and in closely related contexts, it also becomes a technique used by scientists and engineers for acquiring new knowledge. In an early overview of simulation as a productive technique for management scientists, Stanley Vance offered the following definition:

> Simulation means assuming the appearance of something without actually taking on the related reality of that thing. The simulation technique is extremely useful in gaining experience which otherwise could not be had because of cost or technical factors. Among the classic examples of simulation are those dealing with military maneuvers. Without war games it would be difficult and perhaps inconceivable for an army to test its battle effectiveness unless actually engaged in at least a small-scale war . . . The effectiveness of the game depends not only upon the precision in planning but equally as much upon the seriousness with which the group strives to make the simulation approximate the reality.[3]

Much the same might be said about the effectiveness of simulation in engineering and scientific research. To be useful, it must be taken seriously, and to be taken seriously, it must have the appearance of veridicality. Today, the term simulation tends to be employed rather more narrowly, where—especially in the scientific and philosophical literature—it has come to imply the use of digital computers. And these of course no one of Leduc's generation had. Nevertheless, here, in its use as an effective synonym for computer simulation, the feature that was most essential to Leduc—its invocation of a world that is simultaneously artificial and productive—is clearly retained. If anything, it has become even more pronounced.[4]

Uses of computer simulation to study biological systems have exploded over the last decade, and they follow directly from the historical development of simulation in the physical sciences.[5] Indeed, to this day, those who argue for their value are still mainly physical and mathematical scientists. By far the most spirited and highly publicized advocacy comes from those scientists engaged in the venture that Christopher Langton has called artificial life. In Langton's first use of the term, he wrote: "The ultimate goal of the study of artificial life would be to create 'life' in some other medium, ideally a *virtual* medium where the essence of life has been abstracted from the details of its implementation in any particular model. We would like to build *models* that are so life-like that they cease to become models of life and become *examples* of life themselves."[6] One year later, in 1987, Langton organized and convened a conference at Los Alamos with the explicit purpose of inaugurating a new field by that name. Here, he wrote, "Artificial Life (AL) is a relatively new field employing a *synthetic* approach to the study of *life-as-it-could-be*. It views life as a property of the *organization* of matter, rather than a property of the matter which is so organized."[7]

At the time, Langton was a member of the Theoretical Division of the Los Alamos National Laboratory, where virtually all the major techniques of computer simulation had originally been developed. In its earliest incarnation, the term referred to the use of the digital computer to extract approximate solutions from prespecified but analytically intractable differential equations describing the dynamics of thermonuclear processes (such as neutron diffusion, shock waves, and multiplicative or branching reactions). Here, just as in the work of von Dassow and his colleagues (discussed in Chapter 8), what was being simulated was the differential equation itself. However, in an effort to build better theories of fluid behavior, computers soon came to be used to simulate not merely the equations but also the molecular dynamics of real fluids.[8] Many of the computer simulations of biological systems depend on a still further development, namely, on the use of computers to explore phenomena for which neither equations nor any sort of general theory has yet been formulated, and for which only rudimentary indications of the underlying dynamics of interaction are available. In such cases, what is simulated is neither a well-established set of differential equations nor the fundamental physical constituents (or particles) of the system but rather the observed phenomenon as seen in all its complexity—prior to simplification and prior to any attempt to distill or reduce it to its essential dynamics. In this sense, the practice might be described as modeling from above. A method that has proven particularly congenial to this alternative kind of modeling is now associated with the term "cellular automata."

Cellular automata (CA) lend themselves to a variety of uses. In some cases, they are employed to simulate processes for which the equations that do exist are not adequate to describe the phenomena of interest (for example, the emergence of novel patterns in excitable media, turbulence, or earthquakes). In others, CA models might be viewed simply as an exactly computable alternative to conventional differential equations. R. I. G. Hughes stresses the radical difference which the feature of exact computability marks from models traditionally employed in physics, but more noteworthy still is the fact that they are often employed in a different spirit altogether.[9] Here, they are typically aimed at producing recognizable patterns of "interesting" behavior in their global or macro-dynamics rather than in their micro-dynamics. As Stephen Wolfram writes,

> Science has traditionally concentrated on analyzing systems by breaking them down into simple constituent parts. A new form of science is now developing which addresses the problem of how those parts act together to produce the complexity of the whole.
>
> Fundamental to the approach is the investigation of models

> which are as simple as possible in construction, yet capture the essential mathematical features necessary to reproduce the complexity that is seen. CA provide probably the best examples of such models.[10]

Indeed, it is in this last sense that CA find an especially congenial home in efforts to model phenomena for which no equations for the micro-dynamics giving rise to the observed complexity have been formulated. A-Life studies provide a prime case in point.

In truth, however, the project of using the computer to simulate biological processes of reproduction, development, and evolution is of far longer standing than are either of the terms cellular automata or A-Life. Indeed, that project has its origins in the same context (Los Alamos) and in the work of the same people (Ulam, von Neumann, Fermi) from which the basic techniques for computer simulation first arose. If any single individual deserves credit as the father of artificial life, that person is not Langton but John von Neumann.

Cellular Automata and A-Life: A Brief History

Von Neumann's contributions to this field grew directly out of his preoccupations with one of the oldest and most fundamental of all questions about simulation: How closely can a mechanical simulacrum be made to resemble an organism in its most fundamental attributes? What properties would the simulation need to have before it could be said to be alive? In Chapter 7, I focused on the seductive powers of animation, but a long history of self-moving automata had made it abundantly evident that animation by itself was not enough to compel belief in the vitality of a machine. For many, the ultimate test was generation, and for all the ingenuity displayed in the construction of automata, there still remained no plausible rejoinder to the argument of Fontanelle: "Do you say that Beasts are Machines just as Watches are?

Put a Dog Machine and a Bitch Machine side by side, and eventually a third little Machine will be the result, whereas two Watches will lie side by side all their lives without ever producing a third Watch."[11] Thus von Neumann's question: Is it possible to construct an automaton capable of reproducing itself?

Beginning in the 1940s, von Neumann struggled with a kinematic model of automata afloat in a sea of raw materials, but he never fully succeeded in capturing the essential logic of self-reproduction with this model.[12] The breakthrough came with the suggestion of his close colleague, Stanislaw Ulam, that a cellular perspective (similar to what Ulam was using in his Monte Carlo computations)—in which the continuous physical motion required in the kinematic model would be replaced by discrete transfers of information—might provide a more effective approach. Cellular automata, as they have since come to be called, have no relation to biological cells (and indeed, from the beginning, they were also invoked for the analysis of complex hydrodynamic problems), but they did suggest to von Neumann a way of bypassing the problems posed by his kinematic model. Here, all variables (space, time, and dynamical variables) are taken to be discrete. An abstract space is represented as a lattice (or cellular automaton) with a "finite-state" machine located at each node of the lattice. Each such machine evolves in time by reading the states of the neighbors to which it is connected (usually, its nearest neighbors) at time t_n and, according to simple, pre-specified and uniform rules, moving to a new state at time t_{n+1}. Ulam and von Neumann reasoned, and von Neumann soon proved, that the collective dynamics resulting from such simple rules might bear a formal resemblance to the biological process of self-reproduction and evolution.[13]

Von Neumann's initial construction in the early 1950s was cumbersome (requiring 200,000 cells with 29 states for each node), but it made the point. The story of its subsequent development (and dramatic simplification)—from John Conway's Game of Life to Christopher Langton's even simpler self-reproducing "loops" (1984)—has been recounted many times.[14] Somewhat less well known is the history of the use of cellular automata in the modeling of complex physical phenomena (such as phase transitions, turbulence, or crystallization)—an activity that, like artificial life, also exploded in the 1980s.[15] Indeed, the very first conference on cellular automata was also held at Los Alamos (preceding the A-Life conference by four years), and while it provided the occasion for Langton's initial foray into artificial life, the primary focus of the earlier conference was on the physical sciences.[16]

Of paramount importance to the upsurge of interest in CA models in the 1980s was the appearance of a new generation of high-speed parallel-processing computers. Furthermore, that so much of this work has come out of Los Alamos in particular is also no accident, for at that time Los Alamos was one of the few laboratories to have the super-computers that made the execution of CA systems practical.[17] But CA were not simply an extension of conventional modeling practices; they also represented a qualitatively new kind of model.

Cellular automata models are simulations par excellence: they are artificial universes that evolve according to local but uniform rules of interaction that have been pre-specified. Change the initial conditions and you change the history; change the rules of interaction and you change the dynamics. In this sense, the analogy with differential equations is obvious. But equally obvious are many of the differences between CA and DEs. The universe of CA is discrete rather than continuous; its rules of interaction are local and uniform rather than spatially extended and (often) non-uniform; the temporal evolution of a CA system is exactly computable for any specified interactions (given enough time), while DEs are rarely susceptible to exact analytic solutions and only approximately computable when they are not.[18] I suggest, however, that by far the most significant epistemological differences that arise from this new kind of modeling come from the uses to which they are put, the processes by which they are crafted, and the criteria by which they are judged.

Wolfram emphasizes the synthetic aims of CA, arguing that what is of greatest significance is their focus not on the properties of the constituent parts of the system but on "how those parts act together to produce the complexity of the whole."[19] The meaning of synthetic implied here is Kantian—that is, reasoning from principles to a conclusion—and the goal of CA modeling is described as establishing formal similitude between the outcomes of simple algorithmic procedures and the overall behavior of the processes (be they physical, biological, economic, or other) which they are designed to explain. But how, in fact, is the success of these models to be judged? What is the measure of formal similitude? In actual practice, the presentation—and, I argue, the persuasiveness—of CA models of biological systems depends on translating formal similitude into visual similitude. In other words, a good part of the appeal of CA models in biology derives from the exhibition of computational results in forms that exhibit a compelling visual resemblance to the processes they are said to represent.[20] In an important sense, however, such visual portrayals are artifacts—they result from self-conscious efforts to construct visually persuasive representations out of the formal procedures that were designed "to capture the essential mathematical features" of the process. Thus, the persuasive power of CA models depends especially critically on the advances in computer visualization techniques discussed in Chapter 7.[21] Finally, and possibly of greater significance, with the translation from formal to visual resemblance, the nuance of synthetic also undergoes a shift: now, no longer strictly Kantian, the word begins to take on the more mundane (and notably dual) sense of fabrication.

An introduction to the subject of CA by Toffoli and Margolus is instructive on both these points, and I quote from it at length:

> In Greek mythology, the machinery of the universe was the gods themselves . . . In more recent conceptions, the universe is created complete with its operating mechanism: once set in motion, it runs by itself. God sits outside of it and can take delight in watching it.
>
> Cellular automata are stylized, synthetic universes . . . They have their own kind of matter which whirls around in a space and a time of their own. One can think of an astounding variety of them. One can actually construct them, and watch them evolve. As inexperienced creators, we are not likely to get a very interesting universe on our first try; as individuals we may have different ideas of what makes a universe interesting, or of what we might want to do with it. In any case, once we've been shown a cellular-automaton universe we'll want to make one ourselves; once we've made one, we will want to try another one. After having made a few, we'll be able to custom-tailor one for a particular purpose with a certain confidence.
>
> A cellular automata machine is a universe synthesizer. Like an organ, it has keys and stops by which the resources of the instrument can be called into action, combined, and reconfigured. Its color screen is a window through which one can watch the universe that is being "played."[22]

Toffoli and Margolus's comments speak directly to the seductive lure of fabrication, of being able to construct artificial universes of one's own. They also refer to the importance of being able to watch one's creation unfold through the "window" provided by the color screen of a computer monitor. What they do not mention is the power these visual displays also have for the passive viewer, for those in the position not of creating but merely of observing.

The uses of cellular automata for simulating global effects have clearly led to a shift in the meaning of simulation (and, similarly, of model) in scientific research. But they have also encouraged a shift in the meaning of the real. As many have noted, they have come to constitute an "alternate reality"—one that appears ever more easily interchangeable with the real world.[23] The very realism of the visual displays of CA computations plays a crucial role here, and with the dispersion of video recordings in the popular media, it is not just scientists like Toffoli and Margolus who can experience their seductive powers. Furthermore, with the distribution of software for synthesizing look-alike universes, anyone with a computer, a color monitor, and only a modicum of experience can now share in the excitement of "playing God."[24] And in all these uses, the very persuasiveness of the image we see before us on the color screen inevitably generates a degree of uncertainty about its authenticity. Like the character Trudy in Jane Wagner and Lily Tomlin's Broadway hit *The Search for Signs of Intelligent Life in the Universe* (1992), we find ourselves asking: Is the spectacle before us soup or art? Model or reality? Might not these models, as some have argued, give us even closer access to the world as it really is? Might not nature, at its most fundamental, really be constituted of cellular automata?

Questions of just this sort have led G. Y. Vichniac to propose "cellular automata as original models of physics," and to suggest the possibility that the physical world really is a discrete space-time lattice of information bits evolving according to simple rules, an enormous CA running with one of many possible sets of rules.[25] Such a view of the world—as a giant network of digital processors—goes beyond Laplace's dream, but it is not entirely new. It has been advocated by a few mavericks (most notably by Edward Fredkin) ever since the 1960s. On a number of occasions, even Richard Feynman expressed support for the idea.[26] But the notion clearly took on new life in 1983 at the Los Alamos conference on CA, for there it was proposed not only by Vichniac but by others as well (including Wolfram, Margolus, and Toffoli). And at least in the world of computational physics (also referred to as synthetic physics), it has continued to gain legitimacy over the years since.

The most immediately relevant point to note, however, is how short a step it is from such claims about the physical universe to Langton's arguments about biology. Langton would put cellular automata to work to synthesize a universe of living beings, where the ultimate goal would be to create "life" in a new medium. Starting from formal constructions that could serve as models of life, these constructions would soon become so lifelike as to pass from the realm of "as if" over into the realm of the real; they would "become *examples* of life themselves."[27] "Artificial Life," he later reiterated, is the "biology of possible life." It

> is the study of man-made systems that exhibit behaviors characteristic of natural living systems. It complements the traditional biological sciences concerned with the *analysis* of living organisms by attempting to *synthesize* life-like behaviors within computers and other artificial media. By extending the empirical foundation upon which biology is based beyond the carbon-chain life that has evolved on Earth, Artificial Life can contribute to theoretical biology by locating *life-as-we-know-it* within the larger picture of *life-as-it-could-be*.[28]

An interesting ambiguity appears in Langton's expression, for the word "could" might refer to logical or to future conditionality. The primary sense of "life-as-it-could-be" seems to be "life as we could make it be," but this sense oscillates with another—of life as it might now possibly be. Langton's leaning toward the future sense of "could" also marks an important contrast with the use of the same conditional in earlier (and more traditionally mathematical) models. In my discussion of Turing's work in Chapter 3, I argued that the central value of his model of embryogenesis lay in its provision of an in-principle account of how this process could work. But in that context, the primary sense of "could" is strictly logical, and any inference to life-as-we-know-it is limited to historical possibility. If there was an implication of future conditionality in any of the many subsequent uses of Turing's model, it was never made explicit.

Langton's ambitions, by contrast, have been directed to the future from the start. In the short run, however, the prospects for life-as-it-could-be remained largely confined to the world of computer simulations (for Langton, an ideal medium precisely because it allowed abstraction from the constraints of materiality). Over the course of the next few years, attempts to simulate the origin and evolution of living organisms with CA models became a thriving industry with its own conferences, its own journal (*Artificial Life,* founded in 1994), and a cadre of enthusiastic publicists. Moreover, the promise of creating new life forms in a process of "bottom-up" synthesis, in which high-level dynamics and structures emerge spontaneously from the application of simple rules and local interactions, has proven to have immense appeal for audiences extending far beyond the world of computer scientists.[29] Perhaps especially, it proved appealing to readers and viewers who have themselves spent a significant proportion of their real lives inhabiting virtual worlds—as it were, coming of age in cyberspace.

"Genetic Algorithms" and the Evolution of Digital Organisms

N. Katherine Hayles asks, "How is it possible in the late twentieth century to believe, or at least claim to believe, that computer codes are alive? And not only alive, but natural?" To be sure, the design of ever more sophisticated techniques for representing the results of CA models in images that look alive has undoubtedly played a crucial role in establishing the proximity of these models to life-as-we-know-it. What is presented to us is not the specter of replicating codes but the visual depiction of these codes as embodied animal-like forms. Moreover, as Hayles observes, "in these representations, authorial intention, biomorphic interpretation and the program's operations are so interwoven that it is impossible to separate them."[30] Crucial to such interweaving are the powerful effects not only of visual embodiment but also of linguistic embodiment. The A-Life community has developed an extensive biological lexicon for interpreting their models that adds substantively to the sense of proximity to the real-life examples for which they aim.

The principal technique for A-Life simulation goes under the name of genetic algorithms. The method of genetic algorithms (sometimes referred to as adaptive systems) was first introduced in efforts to mimic evolution by natural selection. It exploits the procedures of CA by generating random changes (called mutations) in the population of algorithms (or "genes") with which one begins, procedures for exchanging parts of algorithms (called genetic crossover), and a machine language program that codes for making copies (called reproduction) of the new programs thus constructed. The machine language program (or, as it is often referred to, the body of the digital organism) can be either directly built into the hardware of the computer's central processing unit or stored in memory as data for processing at a later time. To be sure, the actual transformation of these data into a "living organism" requires the activity (or energy) of the CPU, but the important point is that its final form is independent of hardware.

As Thomas Ray explains, "Digital organisms live in the memory of the computer and are powered by the activity of the central processing unit (CPU). Whether the hardware of the CPU and memory is built of silicon chips, vacuum tubes, magnetic cores, or mechanical switches is irrelevant to the digital organism. Digital organisms should be able to take on the same form in any computational hardware."[31] At the same time, however, throughout Ray's discussion (as well as those of others), a persistent ambiguity haunts the relation between genomes (understood here as a collection of genes or algorithms) and bodies, just as it also haunts the relation between information, data, instructions, and programs. For example, Ray writes, "The 'body' of a digital organism is the information pattern in memory that constitutes its machine language program. This information pattern is data, but when it is passed to the CPU, it is interpreted as a series of executable instructions."[32] Yet a few lines down, we learn that "the bit pattern that makes up the program is the body of the organism and at the same time its complete genetic material. Therefore, the machine language defined by the CPU constitutes the genetic language of the digital organism."[33] Here the body of the organism *is* its genetic material.

Hayles makes much the same point. She writes, "These bodies of information are not, as the expression might be taken to imply, phenotypic expressions of informational codes. Rather, the 'creatures' *are* their codes. For them, genotype and phenotype amount to the same thing; the organism is the code, and the code is the organism."[34] Similarly, it is also worth noting that, in Ray's remarks, the code is simultaneously taken as both genome and program, both data and instructions; it is "the bit pattern that makes up the program [that] is the body of the organism" *and,* at the same time, the organism's "complete genetic material" (that is, its genome). In other words, the biological lexicon he employs does not establish but rather presupposes the code as the genome, the genome as the program, and the program as the body of the organism. As Hayles puts it, "Ray's biomorphic namings and interpretations function not so much as an overlay, therefore, as an explication of an intention that was there at the beginning. Analogy is not incidental or belated but central to the program's artifactual design" (p. 150).

One may of course ask: Does such ready assimilation of the words "genes," "programs," and "bodies" in fact matter? In the world of computer science, where terms like "data," "program," and "instructions" are used interchangeably, where "machines" can be built in logic code, and where computers themselves can be virtually embodied, one would have to say that it does not. But in the world of biological science—even one in which computer terminology has so extensively come to inform the literature and where slippage between genomes and organisms (and, equally, between "data" and "instructions") has become chronic—the significance of such distinctions has not yet vanished altogether. For the fact is that, in their actual practices, biologists still live in a world of conventional biological objects. Even with their increasing reliance on computers for visualization and computation, the activity of biological scientists remains grounded in material reality, and in the particular material reality of organisms as we know them.

The immediate issue, in much of this literature is the simulation of evolution by natural selection. How do digital organisms evolve? The universe in which they are said to *live* is defined by the space of the computer's memory and the time required for processing, and "evolution" as the process resulting from their competition for such space and time. Just as is the case for natural selection operating on biological organisms, the winners are those digital organisms with the fastest reproductive rates and, hence, those that occupy the largest share of resources. Ray accordingly concludes, "Evolution will generate adaptations for the more agile access to and the more efficient use of these resources."[35] But here too, just as with biological organisms, the crucial question is: How are the survival and reproductive rates of digital organisms determined? In digital life, the rate at which algorithms ("genes") are copied is specified by a "fitness" function. But the functions specifying the fitness of a digital organism are generally either pre-defined or assigned interactively by the user at each stage of the system's evolution as represented on the screen.[36] In other words, natural selection, as the term is used in this community, depends on the programmer's specification of fitness, either in the initial program or interactively over the course of the program's execution. An obvious question thus arises: In what sense is this evolution by natural selection (at least, as Darwin employed the term)? Would it not be more appropriate to liken the process to artificial selection, as employed by animal and plant breeders?

Notwithstanding such quibbles, the hope has nevertheless persisted that such simulations of natural selection could help us understand the essence of biological processes—processes that must themselves have emerged as a consequence of evolution operating over eons in the natural world. The ultimate challenge in this venture is to explain the evolution of those mechanisms that are required for the development of complex organisms. And here, the very terminology of artificial life encourages the belief that even this problem would not be too difficult. Just as an algorithm is called a gene, the collection of algorithms is referred to as the genome or genotype. As Langton put it, "You can think of the genotype as a collection of little computer programs executing in parallel, one program per gene. When activated, each of these programs enters into the logical fray by competing and cooperating with all the other active programs. And collectively, these interacting programs carry out an overall computation that is the phenotype: the structure that unfolds during an organism's development."[37]

Thus encouraged, A-Lifers have invested a considerable amount of energy over the last decade in the use of genetic algorithms (often in conjunction with "neural nets") to simulate the evolution of developmental mechanisms like those observed in biological organisms.[38] "Emergence" is the operative word here, for it is precisely in their capacity to give rise to global patterns of great complexity that the strength of such models has been seen to lie. Yet despite the proximity to biological processes suggested by all the talk of genomes and programs, the results have so far been disappointing. Tom Ray, for example, has succeeded in establishing conditions in which differentiated creatures can survive but not in which they can evolve: "Replicators in Tierra exhibited only modest evolutionary increases in complexity," and his question is: Why?[39]

Obviously, the failure (at least to date) to generate the kinds of complex mechanisms observed in biological evolution weakens the claim of such models to enhance our understanding of life-as-we-know-it, but because there are other aims to which he can turn, Ray is not discouraged. In fact, he suggests that we might look to organic evolution for "clues as to how we may enhance the richness of digital evolution." Somewhat in tension with his earlier claims, he now writes, "The objective is not to create a digital model of organic life, but rather to use organic life as a model on which to base our better design of digital evolution" (p. 33).

Other efforts to simulate development are sometimes claimed to show more promise. These include, for example, elaborations of an early formal description of the development of simple multicellular organisms that had been proposed by Aristid Lindenmayer, programs for step-by-step elaboration of embryonic networks according to an evolvable set of rules for subdivision, reconnection, and modularization, and a number of others.[40] But here, too, the promise is itself ambiguous. If we are to think of these models as contributing to explanation, what is the explanandum? Is it the architecture and development of organisms or the architecture and development of computers? In a recent review of computer simulations of development, Christian Jacob asks if they can "help accomplish a better understanding of gene-gene interactions, genotype-phenotype mappings, and epigenesis."[41] He continues, however, by citing the application of an "embryonic" technique to the evolution of electronic circuits as one of the more successful efforts in this domain.[42] The question with which he ends is no longer how such models can help us better understand biological dynamics but how they can help us design computers that will behave more like organisms. And somewhat ironically, he concludes that success in "growing" such designs requires us to learn more from nature—especially, it requires taking the most recent lessons from biological research to heart: "Can we interpret genes as a set of instructions? Do genes actually correspond to programs?" Instead of unthinkingly collapsing "instructions" and "data," he urges reexamination of the relations between "program" and "execution." Above all, he writes, "we must rethink our notion that a program and its execution are separable."[43]

Models and Explanation, Causal Agency and Emergence

But what are the promises a successful simulation of biological development is supposed to meet? How is its value to be measured? Of the many ambiguities that haunt the entire endeavor of artificial life, perhaps the most readily accessible are those revealed in responses to such questions as these. Or in responses to the more specific question: How can the growth and development of digital organisms, as models of living forms, contribute toward understanding biological development? Explanation, I have been arguing, depends on the particular needs demanding satisfaction at a particular time and in a particular context. And it is here, in the needs that are to be met, that contemporary studies of artificial life diverge most conspicuously from the concerns of experimental biologists.

The primary commitment of practicing biologists remains to the understanding—and the manipulation—of life-as-we-know-it; and even now, when genetic engineering has become not only a realistic option but a growing industry, interest in life-as-it-could-be extends no further than to the modification of already existing organisms. This difference alone can easily account for the fact that, unlike the computational models discussed in Chapter 8, the models generated by A-Lifers have thus far failed to engage much interest among their biological colleagues, neither among those in academic departments nor among those in industry. The principal point here is not that computer scientists are engineers, and hence dedicated to making new objects, whereas biologists are scientists, interested solely in understanding the world as we find it. For surely, making and understanding are goals common to both domains. Rather, the point is that the instrumental needs of these two disciplines are directed at different targets, or at least, to targets that are still relatively easy to distinguish between. The most immediately pragmatic aims of biological computation are new kinds of computers, whereas those of computational biology are still (and merely) new kinds of organisms. I will argue below that these two ambitions may, in fact, be getting progressively more difficult to distinguish; however, instrumental aims are not the only determinants of explanatory criteria. Scientists in both AL studies and contemporary molecular biology also strive for an understanding of life, and we therefore need also to ask: Are there differences in cognitive aims as well, corresponding to these differences in instrumental aims?

The juxtaposition of Leduc's work at the beginning of the twentieth century with contemporary efforts in AL invites the obvious question of what today's A-life simulations provide that Leduc's earlier simulations did not. Differences in technological promise (both qualitative and quantitative) are obvious, but what about differences in the kinds of cognitive satisfaction these two ventures in synthetic biology have to offer?[44] Surely, these are far less obvious, especially when overshadowed by the commonalities in their basic strategy. Both ventures aim at providing in-principle accounts of the growth and form of living beings by studying analogical dynamics in a different medium. Accordingly, I take as a more interesting question the one that arises when we juxtapose these two ventures with the explanatory goals of experimental developmental biology: Can we identify a qualitative difference in the kinds of cognitive satisfaction that synthetic and experimental biology have, respectively, to offer? So rephrased, the question returns us to a recurring theme—namely, the sorting of strategies of explanation variously described as top down versus bottom up, reductionist versus holist, or analytic versus synthetic.[45]

The writings of Andy Clark provide as good an example as any: Clark reviews the achievements of A-Life simulations in terms of a tripartite sorting of explanatory styles—what he calls homuncular, interactive, and emergent explanations.[46] Homuncular explanations, he explains, are reductionist in that they seek to explain the functioning of a complex system in terms of the behavior of the system's parts; emergent explanations are holist insofar as they are irreducible to the behavior of component parts; and interactive explanations are intermediate between these two extremes—fully compatible with homuncular explanations but, as the name suggests, focusing more directly on interactions between a system and its environment. Although the distinctions Clark wishes to mark are imprecise (he acknowledges, for example, that it is no easy matter to give a precise definition of "emergent"), his categories are familiar, perhaps especially, from their affinity with the categories of dynamical systems theory.[47] Like many others, Clark is particularly interested in those properties that resist more traditional (homuncular) styles of explanation—the kinds of emergent properties exhibited in the modeling of dynamical systems, as in artificial life. To be sure, such properties are not confined either to computer simulations of life or to robotic constructions—they are equally evident in Turing's mathematical model of embryogenesis, or in Leduc's physico-chemical simulations of osmotic growth (although in the latter case, without theoretical description).[48] A-Life simulations clearly exceed these other efforts to model dynamical systems in both their power and versatility, and in that alone can be said to contribute to our cognitive toolkits. But the problem that remains for A-Life is fundamentally the same: it is to tie these fictional imaginings to the real world of homuncular research. The task, Clark argues (and how can one not agree?), "is to develop and carefully *interlock* both types of explanation," and in ways that interactive explanations don't even attempt to do.[49]

In the meantime, however, while we wait to see what such happy couplings might actually look like, progress toward more immediately pragmatic explanatory goals—toward understanding through making—proceeds at a stunning pace.

Creating "Real Life"

In its modern incarnation, use of the term artificial life was at first confined mainly to the world of computer simulations. But when Langton expressed the hope of building models so lifelike that they would be actual examples of life, he deliberately—and provocatively—left open the possibility of constructing these examples in some other (nonvirtual) medium. Indeed, the very ambition to identify "the essence of life" was from the start—for Langton and his colleagues, just as for their precursors in the early part of the last century—linked to the vision of transcending the gap between the living and the non-living. The hope was to create artificial life, not just in cyberspace but in the real world. Rodney Brooks's contribution to the web-based "World Question Center" makes the link explicit: "What is the mathematical essence that distinguishes living from non-living," he asks, "so that we can engineer a transcendence across the current boundaries?"[50] It is hardly surprising, therefore, that artificial life quickly became the operative term referring indiscriminately to digital organisms and to physically embodied robots inhabiting the same four-dimensional world as biological organisms.

In a recent book entitled *Creation: Life and How to Make It,* Steve Grand writes, "Research into artificial life is inspiring a new engineering discipline whose aim is to put life back into technology. Using A-life as an approach to artificial intelligence, we are beginning to put souls into previously lifeless machines . . . The third great age of technology is about to start. This is the Biological Age, in which machine and *synthetic* organism merge."[51] Here, the word synthetic reveals yet another ambiguity, referring simultaneously to artificial structures created in the "mirror-world" of cyberspace, and to those built by engineers, "working with [material] objects and combining them to make new structures" (p. 83). Yet, for all the realism with which digital organisms may be represented on the screen, and for all the seductiveness of the biological lexicon attached to these simulations, engineers, if they are to succeed with such a task, must still grapple with the difference between cyberspace and real space, and with the formidable difficulties encountered in attempting to bridge that gap. Within the A-Life community, however, where explanatory goals remain more abstract, such difficulties tend to be given only glancing attention. As Howard Pattee noted in the first conference held on the subject, "Very little has been said . . . about how we would distinguish computer simulations from realizations of life" (p. 63).

Pattee asserts "a categorical difference between the concept of a realization that is a literal, substantial replacement, and the concept of simulation that is a metaphorical representation." Simulations, he writes, "are in the category of symbolic forms, not material substances" (p. 68). And he reminds his readers of the warning von Neumann himself had issued when he wrote, "By axiomatizing automata in this manner, one has thrown half the problem out the window and it may be the more important half."[52] As Pattee sees it, the problem lies first and foremost in the fundamental relation between symbol and matter, and it shows up with particular urgency for this project in the intrinsic dependence of the "reality" of the organism on the "reality" of its environment.

Yet synthetic life forms, made from material components and assembled in real space, are clearly being built, and in ways that draw directly from work on lifelike simulations in cyberspace. Engineering is a science that specializes in negotiating the gap between symbol and matter, and robotic engineers, like their colleagues in allied disciplines, have well-developed techniques for translating from one domain to the other, for realizing the metaphors of simulation in the construction of material objects. In one sense, computer simulations of biological organisms obviously are, as Pattee writes, "metaphorical representations," but they are also models in the time-honored sense of guides or blueprints. In the hands of skillful engineers, they can be, and are, used as guides to construction in altogether different mediums. Here, the simulated organisms of cyberspace are used to guide the synthesis of material objects mimicking the behavior of biological organisms in real space. But can we take such a physically realized synthetic creature to be a "literal, substantial replacement" of the creature it has been designed to mimic? If it walks like a duck and quacks like a duck, is it a duck? For that matter, does it even meet the less demanding criteria that would qualify it as *alive?*

These of course are questions that worry many philosophers, just as they do the rest of us. And while they are not directly germane to the concerns of this book, like most people, I have some views on the matter. Very briefly, I would argue that even though synthetic organisms in physical space-time are no longer computer simulations, they are still simulations, albeit in a different medium. Yet I have no confidence in an ineradicable divide between simulation and realization. For one thing, mediums of construction can change, as they surely will. They might even come to so closely resemble the medium in which, and out of which, biological organisms grow that such a divide would no longer be discernible. For another, convergence between simulation and realization, between metaphoric and literal constructions, can also be approached through the manipulation of existing biological materials. For example, computer scientists might come to give up on the project of the *de novo* synthesis of artificial organisms, just as most of today's biological scientists seem to have done. The engineering of novel forms of life in contemporary biology proceeds along altogether different lines, starting not with the raw materials provided in the inanimate world but with the raw materials provided by existing biological organisms. Techniques of genetic modification, cloning, and "directed evolution" have proven so successful for the engineering of biological novelty from parts given to us by biology that the motivation for attempting the synthesis of life *de novo* has all but disappeared.[53] The implications of such successes have not gone unnoticed by computer scientists.

In fact, work aimed at bridging the gap between computers and organisms by exploiting techniques of biological engineering is well under way in a number of computer science departments. Some efforts are aimed at harnessing DNA for conventional computational purposes; in others, researchers have begun to use the techniques of recombinant DNA to build specific gene regulation networks, pre-designed to respond to particular stimuli, into real bacteria. One example of the latter is part of a larger and far-reaching project that Tom Knight, Jerry Sussman, and Hal Abelson have recently launched at MIT under the name amorphous computing.[54] The motivation for this project is spelled out in a position paper by Knight and Sussman: "Current progress in biology will soon provide us with an understanding of how the code of existing organisms produces their characteristic structure and behavior. As engineers we can take control of this process by inventing codes (and more importantly, by developing automated means for aiding the understanding, construction, and debugging of such codes) to make novel organisms with particular desired properties."[55]

In efforts such as these, no strictures whatever need obtain against using the biochemical machinery that living organisms have themselves evolved. Nor, for that matter, are there any strictures on what is to count as an organism. In fact, in a subsequent paper on the subject, Abelson and Nancy Forbes write, "The ultimate goal of amorphous computing is to draw from biology to help create an entirely new branch of computer science and engineering concerned with orchestrating the use of masses of bulk computational elements to solve problems." In Freeman Dyson's terms, the novel organisms might be green, or gray, or anything in-between.[56]

The bottom line is that with every passing achievement—in biological computing and computational biology—the gap between computers and organisms becomes both ever narrower and more elusive. Thus, the genetic computer of which Davidson and his colleagues speak (Chapter 8) need no longer be seen as just a metaphor or even as just a model. In two quite different domains—in the designing of new kinds of computers and in the modification of existing organisms—they have begun to approach at least some sense of literality. The route by which this convergence is occurring, however, bears little resemblance to the story usually told about scientific metaphor.[57] Here, the convergence is simultaneously material and conceptual, and one can find no residually literal sense in which any of the referents remain fixed. Furthermore, the metaphor itself can be seen to play a substantive (one might even say instrumental) role in bringing its referents together. Metaphors do far more than affect our perception of the world. In addition to directing the attention of researchers, metaphors guide their activities and material manipulations. In these ways—in many different kinds of laboratories (biological, computational, and industrial), in efforts directed toward a wide variety of ends (theoretical and practical, academic and commercial)—the metaphoric assimilation of computers and organisms works toward the literal realization of hybrid ends. And conspicuous among these is the production of material objects that resist the very possibility of parsing the categories of computer and organism.

Surely, the new "creatures" coming out of biological computing and computational biology are real, but the more pressing question seems to be whether they are *alive.* Has the gap between the living and the non-living, between organism and machine, already been bridged? And if not yet, must we reconcile ourselves to the inevitable joining of these two realms in the near future? If so, how soon? How closely, and in what ways, must the new kinds of entities resemble the products of biological evolution to qualify for the designation "life"?

Such questions are as troubling as they are compelling, and at least part of what makes them so is the anxiety they generate: How will these new creatures threaten our own status on earth? Are we really in danger of being replaced, outpaced in the evolutionary race of the future by a new kind of species?[58] Where, apart from science fiction, are we to look for answers to such questions? Marc Lange argues that the importance of the distinction between living and non-living things is "an empirical question. The answer is for science to discover."[59] Alternatively, Steven Levy claims it is a question for technology, that "by making life we may finally know what life is."[60]

I suggest, however, that it is a mistake to look either to science or to technology. In fact, there are peculiarities to these questions that might disqualify them as having any place at all in the realm of science. For example, when people ordinarily ask if something is alive, the object at issue is already assumed to belong to the biological realm. The question is thus a diachronic one: Is the object (now, organism) either still or yet alive? Has its life ended, or has it begun? Here, however, the question is aimed not diachronically but synchronically: How is this object to be taxonomically classified? Is it to be grouped with the living or with the non-living? But the very asking of the question in this form depends on a prior assumption—namely, that a defining, essential property for the category of life objectively exists, or that life is what philosophers call a "natural kind." Is life in fact a natural kind, and not merely a human kind? Is it not the case, as Foucault so provocatively argued, that the demarcation between life and non-life ought better to be viewed as a product of human than of evolutionary history?[61] Did not the very notion that it is possible to find "a true definition of life" begin only two centuries ago, with the advocacy of Jean-Baptiste Lamarck?[62]

François Jacob, following Foucault, is one of many who believe that it did. He claims that, prior to the nineteenth century, "The concept of life did not exist." What does he mean by this? Clearly not that the term "life" had not been used earlier, for he proceeds by opposing his concept to such notions of life as had already been in use: he writes that his claim is "shown by the definition in the *Grande Encyclopédie,* an almost self-evident truth: life 'is the opposite of death.'"[63] Jacob's complaint with earlier definitions is that they are not constitutive; they do not provide us with a positive characterization of "the properties of living organisms"; they do not tell us what life *is.*[64] Lamarck had stated the problem in similar terms: "A study of the phenomena resulting from the existence of life in a body provides no definition of life, and shows nothing more than objects that life itself has produced."[65] But in order for a characterization to tell us what life *is,* it must presuppose a modal (and structural) essence of life, a defining property of living beings that is not in itself alive but nonetheless absent from all non-living things.

Just as Lamarck's did, Jacob's concept of life depends on a particular taxonomy of natural objects—one which singles out the boundary between living and non-living as primary and relegates all other distinctions between different modes of living to insignificance. This of course was the taxonomy which Lamarck had so strenuously urged at the beginning of the nineteenth-century and which Foucault credited with marking the beginnings of "biology." It contrasts the living not with the dead but with the "inorganic."[66] It highlights one distinction at the expense of others—submerging not only earlier boundaries (most notably, between plants and animals) but also differences of kind between the various sorts of structures that were subsequently to come into prominence (genes, gametes, cells, tissues, organisms, and perhaps even auto-catalytic systems and cellular automata). As long as they possessed the essential defining characteristic, all these structures could—as it were, equally—qualify as instantiations of life.

But by far the most interesting feature of the quest for the defining essence of life, and surely its greatest peculiarity, is that, even while focusing attention on the boundary between living and non-living, emphasizing both the clarity and importance of that divide, this quest for life's essence simultaneously works toward its dissolution. Rodney Brooks and his colleagues are not the first to link the question of what life is with the ambition to transcend current boundaries—the same duality of impulse can already be seen in the writings of Lamarck, and indeed it might be said to inhere in the very demarcation of biology as a separate science. For Lamarck, biology was to be "an enquiry into the physical causes which give rise to the phenomena of life."[67] "Nature has no need for special laws," he wrote, "those which generally control all bodies are perfectly sufficient for the purpose."

Why then demarcate biology as a separate science in the first place? Where, if not to its causes or laws, should we look for the properties that so critically and decisively distinguish the subject matter of this science from that of the physical sciences, that account for the "immense difference," the "radical hiatus" between inorganic and living bodies? (p. 194). Lamarck's answer was to look to the "organization" of living matter: "it is in the simplest of all organisations that we should open our inquiry as to what life actually consists of, what are the conditions necessary for its existence, and from what source it derives the special force which stimulates the movements called vital" (p. 185). Yet, because of his commitment to the adequacy of physical causes and laws, he believed that organization too must have—and must have had—purely physical origins. Hence his interest in spontaneous generation and the origin of life. In other words, just as for many of his nineteenth- and twentieth-century counterparts, the very demarcation of life as a separate domain served Lamarck as an impetus for the breaching of that boundary—if not practically, then at least conceptually.[68] Those who are currently most interested in the distinguishing properties of organization, however, tend to focus more on the construction of material bridges. But either way, conceptually or materially, such bridges invite the formation of new groupings—groupings that necessarily violate older taxonomies. Instead of linking together in a single category plants and animals, they might conjoin computers and organisms; thunderstorms, people, and umbrellas; or animals, armies, and vending machines.[69]

Should we call these newly formed categories by the name of "life"? Well, that depends. It depends on our local needs and interests, on our estimates of the costs and benefits of doing so, and also, of course, on our larger cultural and historical location. The notion of doing so would have seemed absurd to people living not so long ago—indeed, it seems absurd to me now. But that does not mean either that we will not, or that we should not. It only means that the question "What is life?" is a historical question, answerable only in terms of the categories by which we as human actors choose to abide, the differences that we as human actors choose to honor, and not in either logical, scientific, or technical terms. It is in this sense that the category of life is a human rather than a natural kind. Not unlike explanation.

Conclusion: Understanding Development

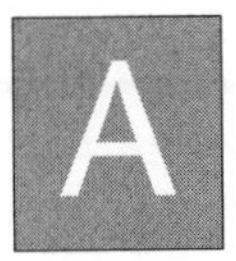

s scientists in the modern era, we generally proceed under the assumption that phenomena, if they are natural, are ipso facto explicable—obliged, as it were, to make sense to us. Friedrich von Schiller's protest that "natural necessity has entered into no compact with man" is relegated to an outmoded era of German romanticism, seen as a last stand against the triumphal progress of scientific enlightenment.[1] The prevailing assumption over the last two hundred years has been that only divine intervention would be capable of releasing the world from its obligation to make sense to us and, hence, that behind all such protests must lie at least the tacit presupposition of a non-natural cause.

But by what mandate is the world obliged to make sense to us? Is such an assumption even plausible? I would say no, and on a priori grounds. One need invoke neither divine intervention nor unknown forces in order to doubt our ability to make rational sense of all that we encounter in the natural world. The human mind does not encompass the world; rather, it is itself a part of that world, and no amount of self-reflection provides an escape from that limitation. Most of us would agree that the mind—along with its capacity to make rational sense—is itself a biological phenomenon and hence a product of evolution, brought into existence by the forces of natural selection. The selective advantages accruing from the ability to make sense of one's immediate and even remote environment are obvious. Yet an evolutionary process that could give rise to a mental apparatus with an *unlimited* capacity for making sense, however desirable such a capacity might seem, is difficult to imagine. And conceiving of a process that would produce minds capable of fully comprehending themselves is more difficult still. I can imagine neither a design nor an associated selective advantage that would be adequate to such a capacity.

Once we grant that our access to the natural world is not unlimited, we might also recognize that other biological marvels crafted by selection could be equally elusive. I see nothing counterintuitive in the possibility that there are phenomena in the natural world extending beyond the grasp of human comprehension—if only by virtue of their sheer complexity. Embryonic development might very well be one of these.

What would it mean to understand development? "Understanding" is a notoriously unstable word, and a central aim of this book has been to illustrate this instability. Nevertheless, the question of whether we do (or can) understand development does often arise.[2] In some contexts, understanding means providing a "reductionist" account—one that invokes only lower order entities. In others, it means providing a program (or algorithm) for "computing" the embryo. And sometimes understanding means both at the same time, as if the two were equivalent. In each case, questions abound.

For example, to what, in the reductionist's account, is development to be reduced? If the goal is the reduction of developmental processes either to a catalogue of genes or to the genome sequence, the possibility that such an account could be fully explanatory has been severely challenged by recent work in the field. Even the more modest claim that development can be reduced to "a total description of the fertilized egg" is probably not valid for organisms whose development depends on a multitude of specific cues from the embryonic environment.[3] We need also to ask: Is such a total description in fact feasible? And even supposing it were, what would it take to move from description to understanding, to "make sense" of such a vast compendium of data? A process can be perfectly rational and in total consonance with the laws of physics and chemistry and yet still not make sense to us, whether because of the finite capacities of our minds or because of our current criteria for making sense.

Recall Gregory Wray's surprise at discovering "a promoter that operates in a logical manner."[4] His point was surely not that we might have expected promoters to behave chaotically, for such a supposition would be manifestly at odds with the reliability with which embryos routinely develop. Rather, his surprise is that the logic of this particular operation *can* be expressed in a form that is accessible to the human mind. The need for understanding, as for explanation, is a human need, and one that can be satisfied only within the constraints that human inquirers bring with them.

The question of whether the egg is (or will be) "computable" harbors a similar ambiguity. If, as Alex Rosenberg suggests, the criterion of computability is satisfied when a process can be executed by a machine, embryogenesis is computable in a trivial sense: an organism is built out of a fertilized egg by a system that we have already agreed to call a machine.[5] The difficulty arises in the conjunction between computing and understanding. The subject of the verb "compute" in this argument is the actual system (that is, the fertilized egg itself, embedded either in an appropriate in vitro environment or within the larger system of the gestating mother) and not the scientist (or reader) who is the presumed subject of the verb "understand." Minimally, then, the equation between understanding and computability requires a narrower definition of "computable" than Rosenberg gives it: the machine that executes the process must be man-made. In fact, this implication is already signaled by Wolpert's use of the future tense ("Will the egg be computable?"), tacitly acknowledging the obvious fact that it is not at present possible to "compute" the embryo from the egg. What then can be said about future prospects?

Whether we will ever be able to execute such a computation depends on many things—not only on whether we will have sufficient information about the processes and initial conditions of development but also on how much time, energy, and money we are willing to invest and on what we hope to gain by such an achievement. Suppose, for example, that the information required is computationally irreducible. By computationally irreducible I mean that no representation less complex than the fertilized egg itself (or more accurately, no less complex than the entire gestational system) would suffice. It is at least plausible to suppose that evolutionary processes have fashioned just such a minimal system; and if so, the obvious question arises: Why bother to duplicate the effort? Given that we already have a machine that executes the developmental computation routinely (namely, the organism itself), why go to all the trouble and expense of building another? The answer to that question depends on what we hope to gain by the effort.

One of the most important goals commonly claimed for scientific explanation in general, and for computation in particular, is predictability. But organisms seen as naturally crafted computational machines can be just as effectively employed for making predictions as can artificially crafted machines, and this is just how they are being used in many contemporary research endeavors. Another goal often asserted is the ability to design and construct new kinds of entities. Yet here, too, real organisms can serve at least as well. The information and technical facility that molecular biologists have acquired over the past several decades make it possible to employ existing natural organisms in the design and creation of new kinds of organisms, genetically modified to perform particular functions and even to conform to particular design principles.

Often, however, we want more: we want the sense that we have understood a process or phenomenon, the feeling that we have brought it within our conceptual grasp. And for many, it is just this sense of cognitive mastery that computability has traditionally promised. But isn't that very expectation premised on computation as the action of a human mind, as the execution of a thought experiment? Isn't our sense of conceptual grasp the consequence of realizing the outcome of a process by means of our own cognitive (computational) activity? We have come to think of Newton's laws as giving us cognitive mastery, but surely that confidence derives in part from the abundance of examples for which we can analyze the relevant equations with our native resources, relying on nothing more than our own mental apparatus. If so, it is hardly a small matter that computations of complex processes are no longer carried out by human subjects. Human cognition is so poorly equipped to match either the speed or complexity of such computations that the role of the human mind has been relegated to little more than that of an observer. What kind of understanding does computability, in the contemporary sense of the term, in fact provide?

One may ask a related question: Once we ourselves no longer perform the computation, does it matter what kind of machine does? Indeed, is it at all obvious that we gain a greater conceptual grasp of a process when we have succeeded in building a machine capable of computing that process than when we can observe the same computation being executed by a machine that has been crafted by evolution?

I raise these questions not in the hope of simple answers but rather to underscore the particular challenges that highly complex processes can pose for the meaning of understanding. As I have repeatedly argued, what leads a student of such processes to say *Aha!* depends on his or her expectations, needs, and desires. For some, the fact that the machines executing a process are of our own design and construction does suffice to yield satisfaction, while others, with more traditional cognitive goals, may hold out for principles they can wrap their minds around, on the model of Newton's laws. These scientists are likely to remain unsatisfied by computability as such, whether the computation is performed by artificial or by natural machines, even if they recognize that they may remain forever frustrated by the very complexity of the processes they seek to understand. Still, others may find their need for understanding satisfied by the possibility of watching the process unfold in intimate detail, or they may be content with global narratives of the kind discussed in Part Two, especially when that narrative is supplemented with concrete handles for manipulating local effects.

The central concern of this book has been with the de facto multiplicity of explanatory styles in scientific practice, reflecting the manifest diversity of epistemological goals which researchers bring to their task. But I also want to argue that the investigation of processes as inherently complex as biological development may in fact require such diversity. Explanatory pluralism, I suggest, is now not simply a reflection of differences in epistemological cultures but a positive virtue in itself, representing our best chance of coming to terms with the world around us.

Readers familiar with recent literature in the history and philosophy of science will recognize a clear affinity between these claims and those of John Dupré, Nancy Cartwright, and others who have been arguing for the fundamental and inescapable disunity of science.[6] But there is also a crucial difference. Dupré grounds his denial that science could ever come to constitute a single, unified project in a claim about how the world actually is. His thesis is "that the disunity of science is not merely an unfortunate consequence of our limited computational or other cognitive capacities, but rather reflects accurately the underlying ontological complexity of the world, the disorder of things."[7] In a similar vein, Cartwright's argument for the inevitably patchwork character of the laws describing our world starts with the premise that "we live in a dappled world, a world rich in different things, with different natures, behaving in different ways."[8] By comparison, the claims I put forth here are about the nature of the scientific pursuit and the essential diversity of interests that drive that pursuit. To the extent that I make a claim for how the world actually is, that claim is only for its irreducible complexity, not for an underlying incoherence. I start from the assumption that the world is devious rather than dappled, too complex to fit neatly into any of our models, theories, or explanations rather than ontologically disordered—that everything is connected to everything else, even if in ways which often elude us and may in fact remain forever beyond our grasp.

My argument for the disunity of science (or, more properly, of the sciences) is thus epistemological and methodological rather than ontological, grounded on the one hand in the disunity of human interests and on the other in the limits of the computational and cognitive capacities inherent in the human condition. If computational limits may frustrate the desire to encompass the world in a unified theory, they simultaneously bring to the fore a variety of other desires. The passion to know—so often claimed as the motor of scientific inquiry—takes many forms, and inevitably so. We seek not only cognitive power, wishing to "gird the sphere" within our imaginative grasp, nor only the kinds of social, economic, and technological power that also come with scientific achievement, but also the more intimate rewards of connection, of wonder, or simply of narrative closure. Furthermore, needs vary—not only from individual to individual but also with subject, historical and disciplinary context, and the social and institutional demands to which the scientific enterprise is inevitably bound.

The question of what qualifies as a scientific explanation may not be answerable in absolute terms, but perhaps—and here I rejoin Cartwright and Dupré—that is only as it should be. For, like other questions about scientific achievement, it leaves the matter open to negotiation in terms of the particular human needs that are, after all, the *raison d'être* of the entire pursuit.

Notes
References
Index

Notes

Introduction

1. As I use the term, epistemological culture differs sharply from Karen Knorr Cetina's closely related concept of "epistemic culture" (1999). Where Cetina aims at a sociological/anthropological description of the "knowledge cultures of science" (for example, high energy physics and molecular biology), I aim at a description of the epistemological assumptions of these cultures. In Cetina's description of her project, "epistemic cultures and sociology of knowledge are two separate issues" (p. 257); by that account, my project would have to be said to belong to the latter. In my usage, epistemological culture bears a considerably closer kinship to what Ian Hacking (*à la* A. C. Crombie's "styles of scientific thinking" [1994]) calls "style of reasoning" (Hacking, 1992). Or, to invoke an even more directly kindred term, this book might be read as one possible response to Lorraine Daston's (1991) call for a "historical epistemology" (see also Tiles and Tiles, 1993).
2. See, for example, Kitcher and Salmon (1989).
3. Weinberg (2001).
4. As such, Part Two might be read as an extension of Mary Hesse's work on "The Explanatory Force of Metaphor." Hesse (1980).

Part One: Models

1. Rashevsky (1960 [1938]), p. vii.

1. Synthetic Biology and the Origin of Living Form

1. Treviranus (1802), for example, wrote, "The object of our research will be the different forms and phenomena of life, the conditions and laws of their existence as well as the causes that determine them. The science which is concerned with these objects, we designate under the name *Biology* or *Science of Life*" (p. 4). Schiller points out that the term had in fact already been invoked two years earlier in England, in a marginal note of a medical treatise by K. F. Burdach (Schiller, 1978), p. 1.
2. Lamarck (1963), p. 282.
3. Loeb (1912), pp. 7–8. Loeb was not the only biologist to conjoin "understanding" with "construction" when writing about the origin of life, nor was such a conjunction confined to the early decades of the century. J. B. S. Haldane (1940), for example, wrote: "Many people are content . . . to say that the origin of life is a mystery beyond the range of science. This may prove true. Some scientists think so. But others are not so modest. They say that if life once originated from dead matter it ought to be possible to repeat the conditions, and make life in the laboratory. If they fail, that will be a triumph for believers in tradition. It will show that some things are beyond human power" (p. 24).
4. It is not this link, however, which dominates Loeb's own work, but a second link, namely, between the origin of life on earth and the beginnings of life of an individual organism. This second link is more difficult to fathom (but see below and fn 15).
5. This problem Loeb (1912) believed he had in fact solved. He wrote, "Although we are not yet able to state how life originated in general, another, more modest problem, has been solved, that is, how the egg is caused by the sperm to develop into a new individual" (pp. 7–8).
6. Some of the major contributors to "synthetic biology" are in the table on the facing page.
7. Still other terms, for example, "morphogeny" and "physiogeny," were also sometimes used; see Leduc (1911), p. 122.
8. Haeckel (1892), chap. 1, p. 414.
9. Herrera (1928), p. 82.
10. Langton (1988), p. 1.
11. Gruenberg (1911a,b).
12. Gruenberg (1911a), p. 231.
13. Gruenberg (1911b), p. 272. Jon Turney (1998) observes that, in the same year, Alexis Carrel's efforts—on the one hand, to construct a

Major contributors to "synthetic biology."

Name	Dates	Field	Home Institution	Major Publications
Traube, Moritz	1826–1895	Physiological chemistry	Akad. Berlin	Production of first "artificial cell" (*Archiv. f. Anatomie u. wissenschaft-liche Medezin*, 1867) (According to Leduc and others, this "should have been the starting point of synthetic biology" [Leduc, 1911: 115])
Montgomery, Edmund	1835–1911	Cell biology	St. Thomas's Hospital	*On the Formation of So-Called Cells in Animal Bodies*, 1867
Rhumbler, Ludwig	1864–1924	Biophysics	U. of Vienna	*Das Protoplasma als physikalisches*, 1913
Benedikt, Moriz	1835–1920	Biomechanics	U. of Vienna	*Krystallization und Morphogenesis*, 1904
Quincke, Georg Hermann	1834–1924	Physics	U. of Heidelberg	"Unsichtbare Flüssigkeitschichten," 1902
Monnier, Denis	1834–1899	Physiological chemistry	U. of Geneva	"Formes des Éléments Organiques," 1882 (with Karl Vogt)
Lehmann, Otto	1855–1922	Physical chemistry	U. of Karlsrühe	*Flüssige Krystalle und die Theorien des Lebens*, 1913
Felix, Jules	1839–1912	Biophysics	U. of Brussels	*La vie des minéraux. La plasmogenèse et le bio-mécanisme universel*, 1910
Leduc, Stéphane	1853–1922	Biophysics	Nantes Medical School	*Théorie physico-chimique de la vie et générations spontanées*, 1910, et al.
Herrera, Alphonso L.	1868–1942	Physiology	U. of Mexico	*Expériences de Plasmogénie*, 1907, et al.
Burke, John Butler	1871–1946	Physics	Cavendish	*The Origin of Life: Its Physical Basis and Definition*, 1906
Lillie, R. S.	1875–1952	Physiology	Clark U.	"The Formation of Structures Resembling Organic Growth . . . ," 1917, et al.

"visceral organism" and on the other hand, to culture beating heart cells in vitro—were widely described in the popular press as yet a third way of "creating life" in the laboratory, although, to my knowledge, the actual term artificial life was not employed in this context (see his chapter 4; see also Landecker, 1999).

14. Gruenberg (1911a), p. 231.
15. This uncertainty seems to me quite revealing, for it suggests the persistence of the ancient view of fertilization as the fusion of a life-bearing active agent (the sperm) and a passive material container (the unfertilized egg). If the sperm is seen as the sole bearer of vitality, that which "causes" the egg to develop into a new individual (see previous footnote), then the simulation of its causal agency by artificial means would indeed appear as an instance of the artificial production of life. For a further discussion of the persistence of this view of fertilization among early twentieth-century biologists, see Keller (1995).
16. Leduc (1911), p. 146.
17. Mann (1948), pp. 19–20.
18. Farley (1974), pp. 162, 214.
19. *Les bases physiques de la vie* (1907), *Les croissances osmotiques et l'origine des êtres vivants* (1909), and *La Biologie synthétique* (1912).
20. Leduc (1911), p. 113.
21. Of this earlier research, Leduc writes, "This remarkable research should have been the starting-point of synthetic biology. The only result, however, was to give rise to numberless objections, and it soon fell into complete oblivion" (ibid., p. 115). Leduc's aim was the rehabilitation and extension of Traube's early triumph.
22. Ibid., pp. 131, 133, 136–137, 139, 140.
23. Ibid., pp. xv, 150, 170. In one of his more extravagant claims, Leduc wrote: "Of all the ordinary physical forces, osmotic pressure and osmosis alone appear to possess this remarkable power of organization and morphogenesis" (p. 131).
24. Responses to Leduc's work were abundant in the Anglo-American press: in the years between 1905 and 1913, at least eight articles on his work appeared in *Scientific American* alone, and after 1911 his books were regularly reviewed in *Science* and *Nature*.
25. Benedikt (1905), pp. 417–418.
26. *Scientific American*, 93 (September 2, 1905): 176.
27. *Scientific American Supplement*, 62 (August 18, 1906).
28. He cites in particular the early work of Moritz Traube and the more recent work of Otto Lehmann. Indeed, he concludes his review with

an extensive quotation from the rather scathing comments of Gaston Bonnier (1907), of the French Academy of Sciences, who argued that Leduc's results were nothing more than "a repetition of Traube's classical experiments" (p. 57).

29. So much so, in fact, that when *Théorie physico-chimique de la vie et générations spontanées* appeared in 1910, Gradenwitz promptly arranged to translate it into German (Gradenwitz, 1912).
30. Gradenwitz (1907), p. 236.
31. Cuénot (1911), pp. 41–42.
32. Cuénot (1913), p. 556.
33. Dean (1911).
34. Gruenberg (1911), p. 237.
35. Gruenberg (1913).
36. Gruenberg (1911), p. 236.
37. Gruenberg (1913), p. 101.
38. *Nature,* May 25, 1911, p. 410.
39. W. A. D., *Nature,* May 15, 1913, p. 270.
40. *Scientific American Supplement,* 76 (August 2, 1913): 77.
41. *Scientific American,* 109 (1913): 82.
42. Bateson (1913), p. 65. I thank Tim Horder for calling my attention to Bateson's interest in mechanical models.
43. Butcher (1911), p. vii. A particularly bitter exchange does indeed appear in the 1907 proceedings of the Académie des Sciences, but the critiques do not quite bear out Butcher's claim. Rather, the main criticism is leveled at Leduc's claim to originality, and at the hyperbole of his interpretations of his results—results based, as Charrin and Goupil write, on "nothing more than appearance" (1907, p. 19). See also Leduc (1907a) and Bonnier (1907).
44. Leduc (1911), p. ix.
45. In his presidential address to the BAAS in 1912, E. A. Schäfer (1912) summed up the current status of the Pasteur-Pouchet-Bastian debate in the English-speaking world as follows: "My esteemed friend Dr. Charlton Bastian is, so far as I am aware, the only scientific man of eminence who still adheres to the old creed, and Dr. Bastian, in spite of numerous experiments and the publication of many books and papers, has not hitherto succeeded in winning over any converts to his opinion" (pp. 11–12).
46. Strick (1999), p. 52. One of the great ironies of this history is that belief in spontaneous generation could be seen simultaneously as evidence of vitalism and of stark materialism. Thus, for example, Bastian (1870) himself could write, "It is the vitalist, however, who

alone has any logical reason for insisting that what may be a good and valid mode of accounting for the origin of crystals cannot be considered to hold good in the case of organisms . . . And if the 'vitalist' wishes to establish the existence of a more fundamental difference between crystals and organisms than we are prepared to grant . . . it remains for him at least to endeavor to show good grounds for the establishment of such a difference" (pp. 174–175). And, in the same year, William Thistleton-Dyer could write in the *Quarterly Journal of Microscopic Science,* "A believer in spontaneous generation is not really an evolutionist, but is only a vitalist minus the supernatural; the special creation which the one assumes is replaced by the fortuitous concourse of atoms of the other" (quoted in Strick, 1999, pp. 76–77).

47. Strick (1999), pp. 51, 52.
48. Huxley (1894 [1870]), p. 236.
49. Mitchell (1911), pp. 64–65.
50. This was the title under which it was immediately published in *Nature,* 90 (September 5, 1912): 7–19, *Science,* 36 (September 6, 1912): 289–312, *Scientific American Supplement,* 74 (October 5–19): 221–223, 226–227, 254–255, and even in the *New York Times,* September 5, 1912, pt. 4, p. 1.
51. See, for example, Pratelle (1913). Schäfer's own collection of press clippings (an archive totaling approximately 440 pages) can be found at the Wellcome Institute for the History of Medicine.
52. Karl Pearson (1900) had earlier made much the same argument: "Spontaneous generation of life could only be perceptually demonstrated by filling in the long terms of a series between the complex forms of inorganic and the simplest forms of organic substance" (p. 350).
53. To support this claim, he cites not only the difficulties of geological and microscopic evidence but also the argument F. J. Allen had made in an address to the BAAS in 1896 that, "owing to the substance being seized and assimilated by existing organisms," such transitional forms would not now be found in nature even if they were produced; see Schäfer (1912), p. 17.
54. Ibid., pp. 162–164. Schäfer, at least in his presidential address, apparently chose to overlook this slight to his own national hero.
55. Ibid., p. xiii.
56. Ibid., p. xv.
57. *Scientific American Supplement,* 74 (1912): 209.
58. Schäfer (1912), p. 4.

59. Bateson (1913), p. 65.
60. Macfarlane (1918), p. 29.
61. See, for example, Lillie (1917); Lillie and Johnston (1919); Lillie (1922).
62. Lillie (1922), p. 121.
63. Leduc (1928). This article is for the most part a recapitulation of Leduc's earlier work, but it is possible to detect a slight shift in language that may have been stimulated by the growing interest in "catalysis," "autocatalysis," and the "dynamic" character of biochemical reactions. Now, he summarizes his principal point as follows: "life is the function of dynamic centers of force in colloidal and crystalloidal solutions, functioning by means of external stimuli which bring about synthesis and decompositions" (p. 70).
64. Buettner (1938), p. 150.
65. Indeed, Oparin wrote, "There is no essential difference between the structure of coagula and that of protoplasm" (*Origin of Life,* quoted in Farley [1974], p. 163).
66. Oparin (1957 [1938]), pp. 56–57.
67. Freud (1950), originally published in 1919.
68. Maxwell (1852), quoted in Cat (2000).

2. Morphology as a Science of Mechanical Forces

1. Thompson (1942), p. 2.
2. A leading researcher working in biomechanics today reports the following comment on a review of a research proposal she submitted to the National Institutes of Health: "The physics of how embryos change shape is neither an important nor an interesting question" (Mimi Koehl, personal communication, October 1999). An indication of the low profile of mathematical models in contemporary biology can be seen in the fact that only 0.04 percent of the U.S. federal budget for biological research in the year 1983 went to support projects in mathematical modeling (including those in population biology); see National Research Council (NRC) report, "Models for Biological Research" (1985), p. 42. See also Israel (1993), Kingsland (1985), and Provine (1971) for accounts of the history of mathematical models in population biology.
3. For reviews of model organisms in the scientific literature, see the articles in *Science,* 240 (June 10, 1988): 4858, a special issue devoted to the subject. For historical reviews, see Kohler (1994), Clark and Fujimura (1992), and Rader (1998).

4. Or so it is generally assumed. In his historical account of the construction of *Drosophila* as a model organism, Kohler (1994), however, argues for a significantly more complex relation between construction and selection.
5. NRC (1985), p. 16.
6. Leduc (1911), p. 113.
7. Thompson (1942), p. 2.
8. Thompson (1942), p. 337; Thompson (1917), pp. 194–195.
9. Thompson (1942), p. 1026.
10. Needham (1951), p. 79.
11. Bohr (1958 [1932]).
12. Thompson (1942), p. 14.
13. See illustrations in Thompson (1942), pp. 394, 501, 505, 563–564, 601, 978.
14. Olby (1986). Thompson was especially critical of attempts to explain embryonic development on the basis of sub-cellular reproductive structures: "In an earlier age," he wrote, "men sought for the visible embryo even for the *homunculus,* within the reproductive cells; and to this day we scrutinize these cells for visible structure, unable to free ourselves from that old doctrine of 'pre-formation'" (1917, p. 159; 1942, p. 342; and quoted in Olby, 1986, p. 287).
15. Thompson (1942), p. 1024.
16. Here also one finds the clearest expression of Plato's influence on his thinking. As he writes, "Material things, be they living or dead, shew us but a shadow of mathematical perfection" (p. 1030).
17. In his preface to the second edition, Thompson (1942) explained, "I wrote this book in wartime, and its revision has employed me during another war. It gave me solace and occupation, when service was debarred me by my years."
18. Thomson (1917), p. 22.
19. Medawar (1958), p. 232.
20. Hutchison (1948), p. 579.
21. Dobell (1949), p. 612. See also L. L. Whyte, *Aspects of Form,* written for the 1951 Exhibition at the Institute of Contemporary Art in London, an exhibition that was in fact designed as a tribute to D'Arcy Thompson. A similar exhibit was hosted by the Museum of Natural History in Washington, D.C., June 1968; see Ritterbush (1968). Most recently, an exhibit on *Growth and Form* has been mounted by the Wellcome Trust in London (February 22 to May 4, 2001; see www.wellcome.ac.uk/en/1/misexhtwo.html).

22. See also Ghyka (1946) and Kemp (1995).
23. Bonner (1998), p. 1.
24. My search in Medline for these two keywords reveals an overlap of less than 1 percent over the last thirty years.
25. Thompson, quoting Goethe (1983), p. 61.
26. Gould (1976), p. 89.
27. Quoted ibid., p. 81.
28. See, for example, Niklas (1992); Bookstein (1997); and McGhee (1999).
29. One of the very few references I have been able to find in the mainstream literature on developmental biology appears in Lawrence (1992: 155), where Thompson's discussion of coordinate transformations is invoked in the context of Lawrence's own speculations about "shape genes." Lawrence writes, "Rapid evolutionary changes of shape and proportion take us back to the observations of Thompson, who was fond of comparing the forms of related animals and showing how an organ of one species could be transformed into that of another by simple proportionate changes in the different axes." Immediately, however, he adds, "it is not possible to be sure whether these comparisons of crabs and skulls, etc., are facile or fundamental." Alain Prochiantz is another exception. But here, too, it is Thompson's transformation theory that is of primary importance; Prochiantz relegates the bulk of Thompson's work to "the pre-genetics of development"; see Prochiantz (2001). The most notable exception however is J. T. Bonner, whose warm appreciation of Thompson's work is a recurrent theme in his writings (see e.g., Bonner, 2002).
30. Thompson (1942), p. 289.
31. Ibid., p. 284. In fact, the sentence immediately following suggests that Thompson did not consider 'heredity' to have much importance at all. In one of his few explicitly mathematical references, he invokes the explanatory power of the method of the differential calculus, and he quotes Poincaré: "In physics, 'on admit que l'état actuel du monde ne dépend que du passé le plus proche, sans être influencé, pour ainsi dire, par la souvenir d'un passé lointain.' (Poincaré). This is the concept to which the differential equation gives expression; it is the step which Newton took when he left Kepler behind" (p. 284).
32. The discourse of gene action will be discussed in detail in Chapter 4, below.
33. Thompson (1942), p. 284. The contrast is sharpened if we read (as I

think we ought to read) Thompson's "merely depends, somehow" as "could merely."

34. Ibid., pp. 643–644. A far more developed argument of this sort can be found in Cartwright (1999a,b).
35. Thompson (1942), p. 1026.
36. As such, Thompson's accounts fall into the category that Robert Brandon (1990, following William Dray) calls "how-possibly" or "potential" explanations—that is, explanations in which none of the "explanatory premises contradict or conflict with 'known facts,'" and which, should these premises prove true, would become "genuine" explanations (pp. 178–179).
37. See Sally Humphries (1993) for her interesting remarks on the "leading on" function of proof as "an unexamined form of extension by analogy" (p. 20).
38. Tennyson, "In Memoriam A. H. H.," LVI.
39. Thompson (1917), p. 2.
40. Thompson (1942), pp. 1028–1029.

3. Untimely Births of a Mathematical Biology

1. Thompson (1942), p. 1027.
2. The Society for Mathematical Biology was founded by George Karreman, Herbert Landahl, and Anthony Bartholomay, all of whom had been students of Nicolas Rashevsky. Karreman was the first president and Landahl, the second.
3. In a report issued by the Committee on Models for Biomedical Research (1985, National Academy Press), the proportion of NIH funds allotted to mathematical models in 1983 is estimated at $1,172,000, or approximately 0.3 percent of the total budget (p. 46).
4. Virtually all available information on Rashevsky's early life and work comes from the writings of his student, Robert Rosen. According to Rosen (1972), Rashevsky completed a doctorate in theoretical physics at the University of Kiev in 1919, but his participation in the White Navy during the revolution obliged his immediate departure from Russia. He found a job teaching physics first at Robert College in Istanbul in 1920, and in 1921 at the Russian University in Prague. During this time, he published numerous articles (in *Zeitschrift für Physik,* for example) on thermodynamics, relativity, and quantum theory.
5. Rashevsky (1960 [1938]), p. vii.

6. Originally published as a supplement to *Psychometrika*. By the time Rashevsky died, he had acquired ownership of the journal, and his widow transferred this to Herbert Landahl. With the founding of the Society for Mathematical Biology in 1973 (of which Landahl was a co-founder), it was taken as the official journal of the society and renamed the *Bulletin of Mathematical Biology.*
7. The name of the committee was changed in 1947 to avoid confusion with the newly formed Committee on Radiology and Biophysics. See Rosen (1972), p. xii.
8. Rosen attributes the collapse to the university's anger at Rashevsky's refusal to submit to the purge of his program demanded by the House Un-American Activities Committee (p. xiii). But the excitement aroused by Watson and Crick's discoveries were surely also a factor, and a number of physical scientists (Neville Symonds, for one) who had been drawn to his program now turned their attention to the new molecular biology.
9. Conrad (1996), p. 8.
10. Rashevsky (1934), p. 188.
11. Discussion following the paper of Rashevsky (1934), pp. 197–198.
12. The classic site for the mathematical analysis of population growth in this period is Alfred J. Lotka's *Elements of Physical Biology* (originally published in 1924, and reissued in 1956 under the title *Elements of Mathematical Biology*). In fact, Lotka's work deals with a far greater range of problems (including evolution, mind, and consciousness), but it is primarily for its analysis of growth functions and inter-species equilibria that he is remembered. Curiously, both Thompson and Rashevsky refer to him only in passing.
13. Rashevsky (1938), pp. ix–x.
14. His commitments were clearly spelled out in the first issue of the Bulletin he founded in 1939: "Emphasis is put upon the mathematical developments, but a description and discussion of experimental work falls also within the scope of the Bulletin provided that description or discussion is made in close comparison with mathematical developments *contained in the same paper.* Outside the scope of the journal are papers of purely statistical nature or papers concerned only with empirical equations."
15. Rosen, unpublished notes; Conrad (1996).
16. See, for example, Rashevsky (1940). Rosen routinely referred to this work as the "Rashevsky-Turing" theory; see especially his discussion of morphogenesis in Rosen (1970), chap. 7.

17. Hodges (1983), p. 429. The same question, in virtually the identical form, will reappear in the discussion of positional information in Chapter 6.
18. Elsewhere, in a paper co-authored by C. W. Wardlaw, Turing wrote, "Unless we adopt vitalistic and teleological conceptions of living organisms, or make extensive use of the plea that there are important physical laws as yet undiscovered relating to the activities of organic molecules, we must envisage a living organism as a special kind of system to which the general laws of physics and chemistry apply"; quoted in Saunders (1992), p. 45.
19. Hodges (1983), p. 641.
20. In 1952, Hermann Weyl's *Symmetry* was published; *On Growth and Form* was reprinted; Adolf Portmann's *Animal Forms and Patterns* appeared in English translation; also published were the Proceedings of a Symposium on Biochemistry and Structural Basis of Morphogenesis held the previous year in Utrecht, and Lancelot Whyte's, *Aspects of Form.*
21. Turing to Young, February 8, 1951.
22. Turing did consult the botanist C. W. Wardlaw, with whom he wrote (but never published) a joint paper on phyllotaxis. This paper was not published until his collected works appeared in 1992; see Saunders (1992).
23. See Saunders (1993), p. 35.
24. Turing (1952), p. 37.
25. The context makes it clear that Turing's use of "falsification" here is intended in the sense of "misrepresentation" and should not be confused with Popper's use of the term.
26. Note the similarity to the words that D'Arcy Thompson had earlier used to express the same hope.
27. Bénard (1901); Rayleigh (1916). Many years later, it was realized that the expected pattern in buoyancy-driven convection could also be a stripe pattern of convection rolls, in addition to the cellular pattern observed by Bénard. Choosing between different possibilities requires nonlinear analysis of a kind that was not developed until long after Rayleigh's work. Furthermore, the convection observed by Bénard is now understood to be driven by temperature-dependent surface tension forces rather than by buoyancy.
28. Turing (1952), p. 37.
29. Hodges (1983), p. 434.
30. In *The Philosophy of 'As If,'* Vaihinger (1924) argued for the importance of both "real" or "genuine" fictions (that is, fictions that are in-

ternally self-contradictory) and "semi-fictions" (which are merely in contradiction with the world). Writing as a biologist, however, William Morton Wheeler (1929) suggested his "as if" was merely "a polite substitute for the phrase 'lies, damned lies, and statistics'" (p. 98). The models discussed here are clearly "semi-fictions" in Vaihinger's sense of that term, but later, in Chapter 4, I also consider the use of "real" fictions in biological explanation. See Fine (1993) for a gloss of Vaihinger's work.

31. See, for example, Hesse (1966), Wise (1977), Siegel (1992), Cat (2001).
32. 1852, quoted in Cat (2001).
33. Quoted in Siegel (1992), pp. 51–52.
34. Cartwright (1999a), p. 36.
35. Weismann had long before proposed that differentiation might result from a programmatic distribution of the hereditary particles of the germ cells during the course of somatic cell division, but by the 1950s, while not yet proven, the prevailing assumption was that the genetic constitution of all cells is the same. Similarly, while the cytoplasm was recognized to be not quite homogeneous, apart from a relatively small number of advocates of cytoplasmic heredity (see Sapp, 1987), the prevailing assumption at the time was that such inhomogeneities were not causally relevant to development.
36. Discussion following the paper of Rashevsky (1934), p. 201.
37. See Delbrück (1949).
38. Keller (1983), pp. 515–516.
39. Keller and Segel (1970).
40. The prevailing account of slime mold aggregation invoked the hypothesis of special founder cells which could (somehow) form centers of aggregation, but this hypothesis only pushed the question of cause one step back: How did such founder cells arise?
41. Mitchel Resnick (1994) describes similar responses to his own early work on computer models of self-organization, and he refers to the expectations of a "leader" or pre-existing "seed" responsible for organizing the patterns that emerge—so prevalent even among his own colleagues—as "the centralized mindset." He writes: "The fact that *even Marvin Minsky* had this reaction is an indication of the powerful attraction of centralized explanations" (p. 120, italics in original).
42. See Keller (1985) for further discussion.
43. Keller (1983), p. 516.
44. In Keller (1983) I also discuss the attraction that alternative models

positing the pre-existence of "pacemakers" had for many mathematical biologists.

45. The subject he called "theoretical biology" was somewhat broader than what I am here calling mathematical biology, but intended to bear a relation to the latter analogous to that between theoretical and mathematical physics. As he wrote in the preface of volume 1 of *Towards a Theoretical Biology,* "Theoretical Physics is a well-recognized branch of science; Theoretical Biology has not yet become one" (1968, p. i).
46. Subsequently published under the title *Towards a Theoretical Biology* (1968–1974).
47. The anomalies of Waddington's intellectual orientation cost him dearly, particularly in his relations to American embryologists. In 1936 Waddington (together with Joseph Needham) submitted a proposal to the Rockefeller Foundation for an institute of physico-chemical morphology at the Strangeway Laboratories. Although the foundation was initially enthusiastic, criticism of Needham and Waddington's "superficiality" and insufficient grounding in experimental embryology (coming especially from George Streeter and Benjamin Willier) persuaded them to drop the project in 1937 (Rockefeller Archives, 1.1, series 401D, Box 41, Folder 523).
48. Waddington to Turing, September 1952, in Turing's papers, University of Cambridge Archives.
49. Ibid.
50. Waddington (1953), pp. 122–123.
51. On this, most embryologists of the time would have agreed. The weight of Waddington's criticism, however, depended on the question of whether the inhomogeneities known to exist in the egg are of central (causal) importance (that is, on whether or not they can be ignored), and on this question, there was not yet general consensus.
52. Kay (2000).
53. See, for example, the reflections of Max Delbrück (1949a).
54. See, for example, Prigogine and Nicolis (1967); Prigogine and Lefever (1968); and, in relation to biology, Prigogine et al. (1969). See also, Wolpert (1969); Keller and Segel (1970); Gierer and Meinhardt (1972); Wilcox et al. (1973). An application of Turing's model to bristle patterns in *Drosophila* that had appeared somewhat earlier (Maynard Smith and Sondhi, 1961) was largely unknown to this community.
55. Murray (1990), p. 119.

56. The online Science Citation Index lists over 1,500 such articles since 1983. On the cover of a 1995 issue of *Nature* (August 31) the striped pattern of a marine angelfish is depicted with the title "Turing Patterns Come to Life." The cover calls attention to two articles: one by Kondo and Asai (1995), and the other by Meinhardt (1995).
57. For recent reviews, see Maini et al. (1997); Meinhardt and Gierer (2000); and Murray (1991).
58. Rashevsky's *Bulletin of Mathematical Biophysics* was revamped in 1973 as the *Bulletin of Mathematical Biology.* This, together with a number of new journals (*Currents in Modern Biology,* begun in 1967 and renamed *BioSystems* in 1972; *Journal of Mathematical Biology,* formed in 1974; and the *Journal of Theoretical Biology,* originally launched in 1961 but by the early 1970s, notably more mathematical in its content) were the primary outlets.
59. Fred Nijhout, a biologist interested in the development of wing patterns in butterflies, was one of those few and his efforts (since the mid-1980s) to integrate his experimental observations with Turing-type models stand out as a notable exception; see Nijhout (1985; 1991).
60. Although the application of reaction-diffusion equations to temporal patterns in chemical systems found experimental corroboration quite early, most notably, in the Belousov-Zhabotinsky reaction) (see Kopell and Howard, 1973), the search for experimental evidence of stable Turing structures took much longer. The first such evidence came in 1990 from De Kepper's group in Bordeaux; see Castets et al. (1990).
61. The notion of a genetic program relied heavily on Turing's own work on digital computation. Yet in the context of these debates, Turing's name (and more specifically, his model of morphogenesis) had come to exemplify a very different (and even opposing) sort of explanation for development—one that was in fact often invoked in opposition to the presumably all-purpose notion of a genetic program that molecular biologists had embraced. Although the notion of program could (and eventually would) be invoked to describe his model, the rules of such a program would not be found encoded in the DNA but rather in the dynamics of particular chemical reactions coupled with the diffusion of their products. For further discussion, see Chapter 4; also, Keller (2000a), chap. 3.
62. Anyone who has observed encounters between experimental biologists and theoretical physicists (or applied mathematicians) will

surely have noticed the bristling of the biologists when faced with the disciplinary hubris of physicists that is so familiar on their own turf as to go unnoticed. Elsewhere, however (see Keller, 2000b), I have argued that the misunderstanding and frustration typical of such encounters is due not only to hubris but also to deep differences in understanding between the two disciplines regarding, first, the nature of theory and, second, its relation to practice.

63. See Meinhardt (1977; 1982), Kauffman (1981), Goodwin and Kauffman (1990), Lacalli et al. (1988), and Lacalli (1990) for some of the many efforts to model *Drosophila* pattern formation by reaction-diffusion.
64. Maini et al. (1997), p. 3608.
65. For example, on butterfly wings, the skin of the marine angelfish *Pomanacanthus,* leopard spots, alligator and zebra stripes, and some of the patterns of slime mold aggregation.
66. John Maynard Smith (1998) offers a particularly interesting review of his own experience of the early allure and the subsequent disappointments of Turing's model for understanding pattern formation in early embryonic developments.

Part Two: Metaphors

1. See Jordi Cat (2001) and Daniel Siegel (1991).
2. Locke's denunciation of figurative speech is especially well known: "If we would speak of things as they are, we must allow that all the art of rhetoric, besides order and clearness, all the artificial and figurative application of words eloquence hath invented, are for nothing else but to insinuate wrong *ideas,* move the passions, and thereby mislead the judgment, and so indeed are perfect cheats; and therefore however laudable or allowable oratory may render them in harangues and popular addresses, they are certainly, in all discourses that pretend to inform or instruct, wholly to be avoided and, where truth and knowledge are concerned, cannot but be thought a great fault either of the language or person that makes use of them. What and how various they are will be superfluous here to take notice, the books of rhetoric which abound in the world will instruct those who want to be informed; only I cannot but observe how little the preservation and improvement of truth and knowledge is the care and concern of mankind, since the arts of fallacy are endowed and pre-

ferred. It is evident how much men love to deceive and be deceived, since rhetoric, that powerful instrument of error and deceit, has its established professors, is publicly taught, and has always been had in great reputation; and I doubt not but it will be thought great boldness, if not brutality, in me to have said thus much against it. *Eloquence,* like the fair sex, has too prevailing beauties in it to suffer itself ever to be spoken against. And it is in vain to find fault with those arts of deceiving wherein men find pleasure to be deceived" (bk. 3, chap. 10, pp. 105–106). Yet, as Paul de Man (1978) has shown, a strong case can be made for Locke's own arguments being figurative through and through.

3. Hesse (1987), p. 311.
4. See Black (1962).
5. See Gadamer (1975).
6. Hesse (1993), p. 56.
7. In particular, see Keller (2000a). However, I must note an important difference between my arguments there and here. My principal claim in *The Century of the Gene* was that the term gene has now acquired so many different meanings that its continuing usefulness is in doubt, whereas here I argue for the productivity of linguistic imprecision. The question thus arises, when is imprecision productive and when counter-productive? My answer is this: imprecision is productive in the absence of literal meanings, and ceases to be productive either when literal meaning is in manifest conflict with implicit meanings (as happened with "gene action" once the chemical identity of the genetic material was agreed upon) or when two or more different literal meanings have been established (as is now the case today for the gene).

4. Genes, Gene Action, and Genetic Programs

1. Morgan (1926a), p. 26.
2. Morgan (1926b), p. 490.
3. See Keller (1996a) for further discussion.
4. Morgan (1924), p. 728.
5. Brink (1927); Muller (1926).
6. Keller (1994).
7. De Vries (1910 [1889]), p. 13.
8. Quoted in Olby (1974), p. 145.

9. Muller (1951), p. 95.
10. For a review of the history of gene as organism, see Ravin (1977) and Keller (2000a).
11. Contra Driesch, Troland (1914) argued that positing genes as enzymes sufficed to provide the organism with its apparent purposiveness, but he attributed rather remarkable properties to enzymes: "the enzyme," he wrote, "and not the mystical 'entelechy' [should be regarded] as the pilot of life's journey" (p. 133).
12. See, for example, Stern (1955); Hartman and Suskind (1965).
13. The attribution of agency to enzymes reveals a similarly metaphoric process at work in the history of chemistry, for, just as in the case of the concept of affinity, it represents an animation of chemical entities. For an interesting discussion of language in chemistry, see Weininger (1998) and Laszlo (1993). Also, physical forces are often said to act, but here, too, the attribution of action reflects an attribution of animation (see fn 14).
14. *On Poetics,* 1457b24–30. Jacques Derrida (1982 [1971]) has an interesting comment on this passage: "Where has it ever been *seen* that there is the same relation between the sun and its rays as between sowing and seeds? If this analogy imposes itself—and it does—then it is that within language the analogy itself is due to a long and hardly visible chain whose first link is quite difficult to exhibit, and not only for Aristotle. Rather than a metaphor, do we not have here an 'enigma,' a secret narrative, composed of several metaphors, a powerful asyndeton or dissimulated conjunction, whose essential characteristic is 'to describe a fact in an impossible combination of words'" (p. 243).
15. Loup Verlet's (1996) analysis of Newton's gravitational "force," and especially his reading of definition IV, suggests an even closer parallel with gene action: there too, both the entity (gravity) and its effect (attraction) await definition. Thus, Newton defines the "force" of gravity as follows: "An impressed force is an action exerted on a body in order to change its state." Gravitational force does not reside in bodies: it is "impressed." But impressed by whom? By the hand of God, of course. Quoting from Newton's letters to Bentley, Verlet writes: "Seen with human eyes, the idea of such a force is 'either absurd or miraculous' (as Leibniz complained), but it can be embedded in a consistent mathematical formalism, behind which may be contemplated the action of 'the divine Arm' impressing motion on 'inanimate brute matter' . . . Newton explains to Bentley that the

'Agent' who is the cause of all the motions in the sky is 'very well skilled in Mechanicks and Geometry'" (pp. 307, 319). In turn, having likened the force to an action performed by "the divine Arm," and expressed in mathematical language, Newton finds the very existence of gravity assured. Responding directly to Leibniz's denunciation in an afterword in the second edition of the *Principia* (the General Scholium), Newton writes: *"[S]atis est quod gravitas revera existat, et agat secundum leges a nobis expositas, et as corporum cælestium et maris nostri motus omnium sufficiat"* (roughly: "It is enough that gravity really exists, that it acts according to the laws we have expounded, and that it yields all the motions of the heavenly bodies and of our sea"). For a fuller account, see Verlet (1993).

16. The absence of a clear definition of the gene finds a parallel in the history of the concept of "chemical affinity." Tracing this history, Michelle Goupil (1991) writes: "Until the introduction of thermodynamic functions to chemical theory, conceptions of affinity maintained a fuzzy, vague, and especially complex character. Many different meanings found themselves united under a single term, thus making that term polysemic" (p. 317, my translation). But it is also clear that the history of the gene departs from the history of affinity, for, where affinity eventually did acquire a stable and relatively clear definition, the gene did not. In fact, only for a brief time, during the first two decades of molecular biology, can it be said that the gene took on a definite meaning; today, the term has once again become manifestly multi-valent, and many workers in the field have come to despair of the possibility of arriving at a precise definition. See Keller (2000b) for further discussion of the current status of the gene.
17. Ravin (1977), p. 19.
18. *Webster's Seventh New Collegiate* (1967) gives two definitions of neologism: "1: a new word, usage, or expression 2: a meaningless word coined by a psychotic."
19. As Aristotle wrote, "The very nature indeed of a riddle is this, to describe a fact in an impossible combination of words" (*On Poetics,* 145 8a26–27).
20. Keller (1995).
21. Morgan (1934), p. 9.
22. Davidson (1986), p. 11.
23. As discussed at greater length in Keller (2000a), the attention of geneticists did not begin to focus on the variable activity of genes (that is, on gene *activation*) until the late 1950s.

24. See, for example, Oyama (1985); Kay (2000).
25. The term genetic program was used almost simultaneously by Ernst Mayr (1961).
26. Jacob and Monod (1961), p. 354. Monod and Jacob, along with their operon model, will prove important again in Chapter 5 in my discussion of the evolution of arguments about feedback and genetic regulation in the 1950s and 1960s.
27. See Keller (2000a).
28. For further discussion, see Chapter 5, and for a general history of debates on the existence of cytoplasmic genes, see Sapp (1987).
29. Jacob and Monod (1959).
30. Monod and Jacob (1961), p. 394.
31. Of course, as André Lwoff and others were quick to notice, it would be a mistake to suppose that genetic information could actually be measured in the same way as the quantity Shannon had defined as information, for while a single base change might spell death for an organism, it would not alter Shannon's measure at all. See Lwoff (1962), pp. 93–94; also Kay (2000).
32. Jacob ([1970] 1976), pp. 2, 9.
33. The original passage reads: "Il y a comme un dessin préétabli de chaque organe, en sorte que si, considéré isolément, chaque phénomène de l'économie est tributaire des forces générales de la nature, pris dans ses rapports avec les autres, il révèle un lien special, il semble dirigé par quelque guide invisible dans la route qu'il suit et amené dans la place qu'il occupe" (Bernard, p. 51).
34. Jacob ([1970] 1976), p. 4.
35. For further discussion of the genetic program from the perspective of current research, see Keller (2000a).
36. See Keller (1995), chap. 3; (2001); Galison (1994); Edwards (1996).
37. Jacob ([1970] 1976), p. 251.
38. See especially von Neumann's 1949 lecture, printed in von Neumann (1966). The first steps of that resolution came with the rise of connectionism, parallel processors, and neural networks.

5. Taming the Cybernetic Metaphor

1. The long prehistory of control mechanisms has been reviewed by many authors, but anyone sensitive to gender issues will surely be struck by the image invoked by Le Roy Archibald MacColl in his 1945 introduction to the subject: "The control art is an old one. With the broadest definition, it is a very ancient art; for one supposes

that if Adam wished to control Eve's vocal output, he had simple mechanisms, such as a well-balanced club, with which he doubtless brought it down a goodly number of decibels. One of the first control devices of general and important application was the centrifugal governor which James Watt invented, about 1790, to control the speed of his steam engine" (1945, p. vii).

2. For example, the biochemist Bernard D. Davis (1961) offers just such an acknowledgment in his "Opening Address" to the 1961 Cold Spring Harbor Symposium, "Cellular Regulatory Mechanisms," when he writes, "As is reflected in the widespread use of the term 'feedback,' such studies of cellular regulatory mechanisms have been influenced to some extent by concepts that have been developed in communications engineering" (p. 1).
3. For a complementary account of the same period, see Abraham (2001).
4. The same year, yet another biochemist was also to invoke the term feedback. In a lecture he delivered at Freiburg on July 5, 1955, entitled "The Steering of Metabolic Processes," Hans Krebs concluded: "The . . . control mechanisms that have been explored here also possess another kind of order. They belong to the mechanisms that are, in electric circuit technology called back coupling ('feedback'). Control through feedback is a mechanism in which the controlled process itself creates conditions that are unfavorable for the process, and thereby brakes it. The braking, in turn, creates favorable conditions and accelerates the process" (quoted in Holmes, 1995). Citing Norbert Wiener, he continued by noting that control by feedback is "widespread not only in the chemical, 'but also in the more complicated physiological organization' of life" (taken from Holmes, 1995, p. 21).
5. Cohn (1958), p. 458. Like Cohn, Warburton (1955) was also concerned with differentiation. His aim was to augment the earlier term homeostasis that already had a wide usage in physiology, precisely in order to explain "the attainment of a predetermined state from a widely deviant initial one" (p. 129). But Warburton's paper invoked the term feedback in the context of, and for students of, classical physiology. His invocation of feedback is thus modern insofar as he draws upon Wiener, but not insofar as he makes no reference to contemporary work in molecular biology. In turn, his paper seems to have had no impact on that literature, and I therefore do not consider it further here.

6. Nanney (1957), p. 136.
7. "Heredity," he wrote, "in this sense, is a type of homeostasis, similar to physiological homeostasis but implying more, since it includes regulation during protoplasmic increase" (p. 134).
8. This shift is marked with particular clarity in Thomas and D'Ari's recent review of genetic regulation, *Biological Feedback* (1990).
9. Waddington cites early examples described by Lotka (1934); Kostitzin (1937).
10. Waddington (1941).
11. Davis (1961), pp. 8–9.
12. Angela Creager and Jean-Paul Gaudillière (1996) have provided a rich and elegant account of the ways in which the meaning attributed to allostery changed over the next few years, largely as a result of Monod and Changeaux's turn to the use of hemoglobin as a model system. Hemoglobin was useful because it was so well studied, because they had access to it, and because its conformational changes exhibit the same kinetics as at least some examples of end-product inhibition; what it does not share with these latter systems is feedback. Not itself an enzyme, hemoglobin is not directly involved in metabolic regulation. It undergoes a conformational change induced by the cooperative binding of oxygen and iron, and neither of these are end products of a reaction directly involving hemoglobin. Indeed, after 1963, allostery comes to refer primarily to conformational changes induced by cooperative binding and only secondarily, if at all, to feedback inhibition. In the context of the present discussion, it is thus of particular interest to note that almost all the references to feedback or to regulatory circuits found in the writings of Monod and Jacob after 1963 appear in relation to systems of genetic regulation rather than to allostery.
13. Monod and Jacob (1961), p. 391.
14. As Creager and Gaudillière have observed, this renaming also constituted a rewriting of the history. "By changing 'feedback inhibition' to 'allosteric inhibition' Monod became a co-discoverer of the phenomenon, now singular, as well as its namer" (ms., p. 34). Not surprisingly, Umbarger (1961), who regarded himself as the discoverer of feedback inhibition, objected to the new term (p. 401).
15. Hindsight provides a powerful opportunity to review the impact of Jacob and Monod's operon model on studies of cellular regulation. Putting aside for the moment the problem of the ambiguity implicit in genetic regulation, that model has proven overwhelmingly suc-

cessful for the description of many kinds of transcriptional regulation, that is, of regulation of messenger RNA (mRNA) operating at the level of the gene. Today, however, transcriptional regulation is widely recognized as constituting only one step in the regulation of enzyme synthesis in higher organisms, even in *E. coli*. For an early discussion of the limitations of the operon model, see Reznikoff (1972), and for a more current review of regulatory mechanisms, see Kimberly Carr (1994).

16. Monod and Jacob (1961), p. 398.
17. The remarkable popularity of this model in this molecular biology community is surely a reflection of Delbrück's exceptional status in that community.
18. Delbrück's translation, p. 1.
19. For example, Lotka (1934); Kostitzin (1937); Denbigh, Hicks, and Page (1948). Judging from the number of times it has subsequently been "rediscovered," it was apparently also easy to forget, even for mathematical biologists.
20. A very general sketch of such a mechanism had also been suggested in general terms by Wright (1945) when he wrote: "On this view the origin of a given differentiated state of the cell is to be sought in special local conditions that favor certain chains of gene-controlled reactions which cause the array of cytoplasmic constituents to pass the threshold from the previous stable state to the given one" (p. 299).
21. Horowitz and Mitchell (1950), p. 479.
22. Nanney (1957), p. 155.
23. Six years later, Ebert and Wilt (1960) echo Waddington's usage in citing Delbrück's model as "one logical and popular alternative to the unequal distribution of cytoplasmic genetic units" that can explain how "new phenotypic qualities appear and persist (cf. Waddington's canalization hypothesis)" (p. 263).
24. Waddington (1954), p. 117.
25. Gaebler (1956).
26. Cohn (1958), p. 464. Leo Szilard (1960) published a similar model that he had developed and presented in 1957 to account both for induction and repression of β-galactosidase and for end-product inhibition in the biosynthetic pathways.
27. Monod and Jacob (1961), p. 398.
28. Monod and Jacob (1961) also cite Novick and Szilard (1954) as the first experimental observation of allosteric inhibition (p. 390). Creager and Gaudillière (1996) note that Monod and Jacob's citing of

Novick and Szilard (rather than Umbarger) "served Monod's interest" (p. 34); I would add that their citing Delbrück's model as the theoretical precursor to feedback inhibition served the same interest. Both Szilard and Delbrück were important members of Monod's "club," while Umbarger was not.

29. Jacob and Monod (1963), p. 30.
30. Crick (1958).
31. Jacob and Monod (1963), pp. 53, 58, 59.
32. Thomas and D'Ari (1990), p. 2.
33. The shifts in their use of Delbrück's model might also reflect the weakening association of allostery with feedback in Monod and Jacob's minds (see Creager and Gaudillière, 1996). I would suggest, however, that both this shift and the weakening of allostery's tie to feedback can be seen as reflecting their declining interest in feedback operating on any level other than the gene.
34. Moore (1963), pp. 236, 239.
35. The reference here is to Monod's famous assertion, cited in Judson (1979), p. 613.
36. See Keller (2000a) for further discussion of such mechanisms.

6. Positioning Positional Information

1. Stent (1968), p. 390.
2. For example, Wolfgang Beerman in Tübingen, Ernst Hadorn in Zurich, P. D. Nieuwkoop in Utrecht, John Gurdon in Cambridge, England, and Edward B. Lewis in Pasadena.
3. Of particular importance in this renaissance was the pioneering work of Christiane Nüsslein-Volhard and Eric Wieshaus on maternal effects in *Drosophila* embryogenesis, reviewed in Keller (1996).
4. Michael Ashburner (1993), for example, writes: "The introduction, and subsequent democratization, of the technology to clone and manipulate genes has clearly been the most significant event not only for the study of *Drosophila,* but also for biology as a whole, since the discovery of the nature and structure of the genetic material itself. Indeed, it is difficult not to reflect on how important, in general, the introduction of new technologies has been for the advancement of our science in this modern period" (p. 1500).
5. See Wolpert (1970), p. 200.
6. Waddington (1961), p. 69.
7. The concept of "field" had multiple origins. Beloussov credits the Russian embryologist Alexander Gurwitsch (1997), with the first ex-

plicit formulation of "the idea of a 'field' as a supracellular ordering principle governing the fate of cells" in 1912 (p. 773), but De Robertis credits Ross Harrison's 1918 studies of newt forelimb development. Hans Spemann introduced the related notion of "a field of organization" in 1921, and Paul Weiss, his concept of a field in 1923, in discussion of his own experiments on regeneration. The concept of gradient is even older and can be traced back to the work of Abbé Trembley. For further discussion of fields and gradients, see Opitz and Gilbert (1997); Gilbert, Opitz, and Raff (1996); Maienschein (1997).

8. Opitz (1985), p. 1.
9. Wolpert (1968), p. 125. The symposia were held at the Rockefeller Foundation Villa Serbelloni on Lake Como and were officially sponsored by the International Union of Biological Sciences. The vision of a theoretical biology had earlier inspired the Theoretical Biology Club Waddington had been a member of in the 1930s. With fellow members Joseph Needham and the philosopher J. H. Woodger, Waddington had been preoccupied with the particular challenges of developmental biology, and although all three were committed reductionists (in the sense of physical-chemical reductionism), they were at the same time directly concerned with the question of whether these problems required a distinctive kind of theory (that is, distinct from physics). But from these early efforts to formulate a theoretical biology, little now survives.
10. For the original illustration, see Wolpert (1968), fig. 2, p. 130.
11. The conjoining of these two terms (positional information and pattern formation) soon becomes Wolpert's trademark (he published at least seven articles under that title between 1970 and 1994 [1970; 1971; 1972; 1981b; 1985; 1994a; Wolpert and Stein, 1984]). Indeed, so tightly associated do these terms become in Wolpert's own mind that he comes to think that he coined "pattern formation" as well (1989): "I like to think that I invented the term 'pattern formation': I had great difficulty finding a suitable name and even consulted a classicist to see if another word would do. For pattern, as normally used in English, is not quite the right word, the essential connotation being template. Pattern formation does, now, seem to have just the right meaning" (p. 12). In point of historical fact, however, the term appears in the literature with considerable frequency from the early sixties on. In Waddington's (1962) Jessup Lectures (April–May 1961), for example, he wrote, "I propose using the two well-known terms 'morphogenesis' and 'pattern formation' . . . I shall use pattern

formation for processes in which we wish to distinguish different spatial parts within the developing system and to discuss their geometrical relations" (pp. 2–3). Furthermore, Waddington had already used this term in *The Strategy of the Genes* (1939, pp. 193–194).

12. Wolpert (1970), pp. 201, 228. A fuller and noticeably more confident version of Wolpert's argument appears in print a year after its initial presentation; see Wolpert (1969).
13. Wolpert (1989), p. 12. The Friday evening lectures are a long established tradition of the Marine Biological Laboratories at Woods Hole that routinely draws a large and diverse audience of biologists from across the spectrum of the life sciences. Wolpert's lecture on "The Cell in Morphogenesis and Pattern Formation" was given on July 26, 1968.
14. Interview with Keller, June 12, 1994.
15. Susan Bryant, best known as the co-author of a later and alternative model for limb development (French et al., 1976), now recalls, "It was the right approach for people like me—non-mathematical, but theoretically minded" (interview with EFK, February 16, 1996).
16. Smith (2000), p. 85.
17. "30 Years of Positional Information," London, September 1996, organized by Cheryl Tickle and James Smith. In an article published the same year ("One Hundred Years of Positional Information"), Wolpert extends its history further back in time (1996).
18. Zwanziger (1989), p. 134.
19. Public Lecture, California Institute of Technology, spring 1996.
20. In fact Wolpert (1986) credits Hildegard Stumpf's 1966 paper (of this very title) as offering "the first very clear statement of positional information" (p. 356).
21. Wolpert (1969), p. 44.
22. Quoted in Wolpert (1986), p. 347.
23. Wolpert (1969), p. 6. Wolpert (1987) also distinguished his own approach from more recent efforts such as that exemplified in Alan Turing's 1952 paper (see Chapter 3): "Turing's approach to pattern formation was to try and set up a prepattern using reaction-diffusion mechanisms. As such it has nothing to do with the tradition that thought in terms of gradients" (p. 359).
24. Wolpert (1971), p. 184. An even more explicit statement of this position appears in a paper co-authored by Michael Apter (1965). Here the authors write, "Genetic information seems only to mean factors essential for inherited characters," and argue for viewing the egg as "containing the instructions of a programme for development."

Elaborating this argument, they conclude: "If the genes are analogous with the sub-routine, by specifying how particular proteins are to be made . . . , then the cytoplasm might be analogous to the main programme specifying the nature and sequence of operations, combined with the numbers specifying the particular form in which these events are to manifest themselves" (p. 257).

25. Wolpert and Lewis (1975), p. 14.
26. Indeed, by 1996 Wolpert defines the developmental program as a subset of the genetic program: "The genetic program refers to the totality of information provided by the genes, whereas a developmental program may refer only to that part of the genetic program that is controlling a particular group of cells" (p. 21). For further discussion of the relation between developmental and genetic programs, see Keller (2000a).
27. *Drosophila* had historically been the organism of choice for geneticists because of the abundance of phenotypic markers in the adult fly and because of its short generation time, but embryological analysis was inhibited by the fact that the *Drosophila* egg is so small and, at least in its untreated form, so opaque.
28. Ashburner (1993), p. 1501. See Keller (1996) for a discussion of this early work in its historical context.
29. Driever and Nüsslein-Volhard (1988a,b).
30. Approximately fifty such maternal genes have been identified as playing crucial roles in the formation of a normal embryo.
31. The initial localization of *nanos* mRNA plays a correspondingly important role in the specification of posterior structures.
32. Driever and Nüsslein-Volhard (1988b), p. 95. One year later, Nüsslein-Volhard and her colleagues would make an analogous claim for the function of the dorsal protein in the establishment of dorsal polarity (Roth et al., 1989). Here however the gradient is not in the cytoplasmic distribution of dorsal but in the nuclear uptake of the protein.
33. Agutter et al. (2000), pp. 71, 75.
34. For example, retinoic acid in chick limb development, Differentiation-inducing factor and cyclic C-amp in cellular slime mold.
35. The frequency of citations listed in Medline under the term morphogen grows steadily from the early 1970s (when it averages less than one per year) to the present (over 50 per year), but the most precipitous jump occurs between 1988 (12 citations) and 1989 (36 citations).

36. Unlike earlier candidates, the bicoid protein molecule is so large that it would normally not be a candidate for diffusion. But as Driever and Nüsslein-Volhard (1988b) wrote, "In the syncytial blastoderm in early insect embryos no cell boundaries limit the diffusion of proteins the size of the *bcd* molecule" (p. 103). Indeed, for some developmental biologists, this very fact limits the explanatory power of their analysis, and in spite of all the remarkable work of Nüsslein-Volhard and her colleagues, talk of diffusion continues to raise their hackles. *Drosophila,* they remind us, is a rather special case just because of its long pre-cellular (syncytial) stage of development: for most organisms diffusion would be blocked by the presence of cell membranes from early on.
37. Wolpert (1989), p. 3. Here, PI is credited with suggesting possible mechanisms rather than as itself providing the mechanism.
38. The implied reference here is to the observation of a linear sequence of HOX genes, first described by Lewis (1978), and the correlation between the order in which such genes are laid out along the chromosome and the order along the organism's anteroposterior axis in which they are activated; for further explication, see Wolpert et al. (1996).
39. Wolpert (1989), pp. 4, 10.
40. Mullins et al. (1996), p. 81.
41. Wolpert (1986) himself emphasizes the importance of "the transition from thinking about metabolism to thinking about information flow," although he does not elaborate on the strategic nature of its importance (p. 359).
42. Wolpert (1989), p. 4.
43. In his interview with Smith (2000), Wolpert describes a plausible model as one in which "if you put in reasonable numbers, you would see that in principle the thing would work and there are no real holes in it. That would be a solution . . . I'm not interested in detail, I must confess."
44. For a review, see Lawrence (1992).
45. There is some evidence suggesting that, at least initially, the mathematical models of morphogen gradients developed by Alfred Gierer and Hans Meinhardt may have had some influence on Nüsslein-Volhard's approach, but I have found none suggesting that Wolpert's models did. And, ultimately, Nüsslein-Volhard came to disclaim even the influence of Meinhardt's models. In a letter to Keller

on October 4, 1992, she wrote: "I think it is important to note that the influence of Meinhardt's predictions and suggestions on our experiments was always next to zero"; see Keller (1996), pp. 339–340.

46. Lawrence (1992), p. 148.

Part Three: Machines

1. Writing a first draft of this introduction while visiting the École Normale Supérieure in Paris, I found a full page of *Le Monde* (January 24, 2001, p. v, articles by Caterine Tastemain) devoted to the new microscopy of living processes. The take-home message is simply and clearly given in the headline: "To understand biological processes, scientists need to observe the cell in its living state" (my translation).
2. Shapin and Schaffer (1985).
3. Hillis (1993), p. 80.
4. See Fine (1992), as well as the discussion in Chapter 3.
5. A number of historians and philosophers of science (see, for example, Woodward, 2001) have recently argued that we have for too long tended to overlook the pragmatic dimensions that are, and always have been, crucial to the meanings of understanding and explanation in the natural sciences. Yet something is surely changing, and perhaps especially so in biology, where these pragmatic dimensions are now proving so salient they can no longer be ignored.

7. The Visual Culture of Molecular Embryology

1. In Greek, *theôria* originally meant a looking at, or viewing, and *theôreô,* a spectator. For more general discussion of the visual metaphor for knowledge, see Heidegger (1977); Keller and Grontkowski (1983); Keller (1986); Jay (1988); Levin (1993).
2. The same, of course, must be said of the intertwining of seeing and touching (see Hacking 1983; Keller 1996), as well as of the undeniable significance of bodily experience in the acquisition of knowledge (see Polanyi 1962; Johnson 1989; Sibum 1998; Cat 2001). My focus in this chapter on visual access is not intended either to discount or to undermine the insights from recent critiques of ocularcentrism in modern science but rather as an occasion to meditate on the implications of that very tradition for what has counted, and for what continues to count, as explanation.

3. Tufte (1997), pp. 9–10. For a useful overview of the literature on visual representation, see Lynch and Woolgar (1990).
4. See Hacking (1983); Brown (1987).
5. Daston (1991), p. 266.
6. The Oldenburg quotation is in Easlea (1981), p. 85; Leeuwenhoek in Roger (1997), p. 149; and for discussion of Senebier, see Huta (1998).
7. For an example of an explicitly feminist reading, see Keller (1986), where I wrote: "The ferreting out of nature's secrets, understood as the illumination of a female interior, or the tearing of nature's veil, may be seen as expressing one of the most unembarrassedly stereotypic impulses of the scientific project. In this interpretation, the task of scientific enlightenment—the illumination of the reality behind appearances—is an inversion of surface and interior, an interchange between visible and invisible, that effectively routs the last vestiges of archaic subterranean female power" (p. 69).
8. See Berrill (1984), p. 8.
9. The hostilities between Hooke and Newton are legendary among historians of science, and while they were never explicitly posed as a debate over the relative value of observation and logic, echoes of that debate might nonetheless be discerned in the conflicting temperaments of these early pioneers of the microscope and the calculus.
10. Berrill (1984), p. 4.
11. Quoted in Roger (1997), p. 163.
12. A more nearly contemporary echo of this sentiment can be found in Ramón y Cajal's autobiography (1996 [1937]), p. 307, where he wrote of the "vivifying and stimulating power of things seen, that is, of the direct perception of the object, as compared with the very weak, not to say ineffectual influence of the same things when they reach the mind through the cold and second hand description in books."
13. Quoted in Wilson (1995), p. 89.
14. Hooke, p. 186.
15. Roger (1997 [1963]), p. 149.
16. For a lively account of the history of preformationism, see Pinto-Correia (1997).
17. Technical improvements in microscopy surely contributed to this renewed confidence, but as Hacking (1983) has discussed, at least equally important was the concurrent development of techniques for coordinating visual effects with manual manipulations. See also Keller (1996).
18. It was Strasburger's 1879 observation of cell division in living cells

that confirmed the temporal sequence of events that could previously only have been conjectured from the cytological examination of fixed stages; see Hughes (1989), p. 64. Furthermore, even though visibility in living cells was far more limited than in fixed and stained cell slices, observations of in vivo structures were often crucial in establishing the confidence that the entities identified in the latter were not artifacts of staining or fixing. For a nuanced account of the play between theory and observation in the emergence of cell theory, see, e.g., Parnes (2000).

19. Wilson (1925), p. 77.
20. Ramón y Cajal (1996), pp. 526–527.
21. Landecker (1999), p. 63.
22. The fate map of early ascidian development produced by E. G. Conklin in 1905 is a particularly stunning example of how much could be learned with such rudimentary techniques. The magnitude of Conklin's achievements can clearly be seen from Eric Davidson's reproduction of Conklin's drawings, shown in juxtaposition with more recent studies of the same system (1986, fig. 6.3, following p. 256; see also Scott Gilbert's web site, http://zygote.swarthmore.edu/Conklin/Conklin.html).
23. Berrill's (1984) full quote is as follows: "Biology above the molecular level is eminently and inherently visual, and much of its progress during the past two centuries has resulted from the invention of visual aids ranging from simple magnifiers to the scanning electron microscope. The human difficulties relate to scale and time. Microscopic beginnings need to be enlarged, biological happenings may need to be speeded up or slowed down for the eye to catch the movement. Finally, what is discovered has to be portrayed visually to others, generally in four dimensions. This is the problem" (p. 4).
24. See Nick Rasmussen's excellent study of the electron microscope (1997) for an in-depth analysis of the impact of this new technology on the biological sciences.
25. My remarks here refer primarily to the transmission electron microscope. Scanning electron microscopy, introduced into experimental biology somewhat later, has become useful for observing whole mount specimens rather than slices, but it too is restricted to the observation of dead specimens. I thank Greg Davis for calling this distinction to my attention.
26. Albert Szent-Gyorgyi (1972) wrote: "This downward journey through the scale of dimensions has its irony, for in my search for the secret of life, I ended up with atoms and electrons which have no life at all.

Somewhere along the line life has run out of my fingers" (p. xxiv). Just how impossible is the preservation of life under the electron microscope is made dramatically clear by Rasmussen's report that "the electron radiation experienced by a specimen in the microscope beam is said to be about the same as would be received from a ten megaton H-bomb blast 30 yards away" (1997, pp. 26–27).

27. Inoué (1953), p. 499. In fact, Inoué (1993) reports that, even in 1966, when Keith Porter finally succeeded in depicting microtubules in the electron microscope, Porter borrowed Inoué's slide showing the birefringence of spindle fibers in the living cell to prove that his own image was not an artifact (p. 106).
28. Allen (1985), p. 279. Edward L. Chambers makes a similar point in his introduction to Chambers and Chambers (1961), first, concerning the development of cytological techniques in the latter part of the nineteenth century and, second, concerning the electron microscope: "These techniques seemed so fruitful of results that the aim of their originators—to visualize live cell structure—was lost sight of. More and more attention was focused on killed and fixed cytological structures, despite the fact that many of the details, now so clearly visible, might well have been due to the coagulating action of the particular fixative used . . . During the past several decades intense preoccupation with fixed tissues has waned. Recently, however, with the advent of electron microscopy and its fascinating revelations, there have been indications that the cycle may repeat itself!" (p. xxiii).
29. Bohr (1958) described this incompatibility as follows: "In every experiment on living organisms there must remain some uncertainty as regards the physical conditions to which they are subjected, and the idea suggests itself that the minimal freedom we must allow the organism will be just large enough to permit it, so to say, to hide its ultimate secrets from us. On this view, the very existence of life must in biology be considered as an elementary fact, just as in atomic physics the existence of the quantum of action has to be taken as a basic fact that cannot be derived from ordinary mechanical physics" (p. 9).
30. It should be noted that the first reporter genes developed were used with toxic dyes, and hence could only be employed for the observation of fixed embryos.
31. Matus (1999).
32. Whitaker (2000), pp. 181, 180.

33. For a comprehensive review of video microscopy, see Inoué and Spring (1997); for a review of confocal microscopy, see Pawley (1995); and for magnetic resonance imaging, see Blumich and Kuhn (1992).
34. See Inoué (1981); Allen et al. (1981a,b).
35. Typically, the computer also averages the data over some time interval to remove extraneous noise. Such averaging results in increased resolution, but at the cost of losing information about processes occurring within that time interval.
36. Foe et al (2000). These films are now available on-line: http://dev.biologists.org/content/vol127/issue9/images/data/1767/DC1/DEV7765.mov
37. Its invariant pattern of cell lineage makes the roundworm *C. elegans* a notable and invaluable exception.
38. I want to thank Antoine Triller at the École Normale Supérieure for clarification of these points.
39. Beginning with the work of O'Rourke and Fraser (1990) and Cornell-Bell et al. (1990).
40. Quoted in Hoke (1993), p. 19.
41. Service (1999), p. 1668.
42. Miller et al. (1995).
43. Wolpert and Gustafson (1961).
44. McClay (2000).
45. Kulesa and Fraser (1998), p. 327.
46. Kulesa et al. (2000); Kulesa and Fraser (2000).
47. Jacobs et al. (2000).
48. See Engel et al. (1999); Piston (1999); You and Yu (1999).
49. For example, William Mohler suggests that part of the power of computational rendering lies in the fact that "it produces real views of the specimen that can never actually be seen on the microscope and allows the viewer to use stereo vision and a sense of spatial motion to discern the position and trajectory of details" (1999, p. 3061).
50. See Hearst (1990); Orr-Weaver (1995); Monteith (1995); Herschman et al. (2000).
51. Kirschner et al. (2000).
52. Hoke (1993); Service (1999); Whitaker (2000).
53. Alberts (1998), p. 291.
54. Kirschner et al. (2000), p. 87.
55. Mehta et al. (1999), p. 1689.
56. See, for example, the CD-ROM distributed as part of *Trends in Cell Bi-*

ology 9(2) (1999); also, "Molecular Visualization for the Masses—3-D imaging resources for nonstructural biologists"; http://news. bmn.com/hmsbeagle/90/reviews/insitu; www.microscopyu.com; and www.microscopy.fsu.edu.

57. Shapin and Schaffer (1985).
58. Mitman (1999).
59. See Mohler (1999).
60. See Hacking (1983); Bechtel (1994).

8. New Roles for Mathematical and Computational Modeling

The first part of this chapter is taken from Keller (2000d) and the second, from Keller (2000c).

1. See, for example, Buchwald (1995) and Krieger (1992).
2. I have learned from Sunny Auyang (1998) that the same metaphor appears in the writings of the Chinese philosopher Zhuang Zhou, albeit with a difference worth noting. According to Auyang, the Chinese parable attributes the endurance of the knife's sharpness as much to the butcher's skill and discernment as to the existence of universal joints: "The master's knife has been used for nineteen years to operate on thousands of oxen, but its edge is as sharp as it was when it left the grindstone, for the master discerns cavities and crevices and guides the blade through them without resistance" (pp. 85–86).
3. See Hacking (1987) and Keller (1992), Chapter 4.
4. *Webster's Seventh New Collegiate* (1967).
5. See Cartwright et al. (1995); Morrison (1998; 1999); Cartwright (1997; 1999); Morgan and Morrison (1999).
6. I discuss this point further in Keller (2000c).
7. Yuh et al. (1998); von Dassow et al. (2000).
8. Worth noting in particular is Davidson's immensely influential textbook, *Gene Activity in Early Development,* first published in 1968 and now in its third edition.
9. Yuh et al. (1998), p. 1896.
10. Wray (1998), p. 1871. One might well ask about the meaning of "logical" here. Given that the stages of embryonic development are highly predictable, what would it mean for a promoter to behave in an "illogical" manner?
11. Ibid., p. 1872.
12. It should be noted that the actual details of this program are in some tension with its description as a program "directly encoded in the

DNA." While the DNA does indeed provide the source code for the proteins participating in these interactions, the relevant sequences are scattered throughout the genome. Furthermore, the dynamics of interaction between proteins and DNA binding sites are determined by such structural features of the proteins as cannot be predicted from amino acid sequence alone. Indeed, given the extent of splicing, even the latter cannot be fully predicted from the sequence of the source code.

13. Yuh et al. (1994).
14. See Yuh and Davidson (1996).
15. Or more, if the binding factors exist in more than one possible state (which they do) or if the restriction to two states for the binding sites were dropped (as the experimental data in fact demand). The number would be raised by an additional order of magnitude if the temporal order in which these sites are occupied is relevant (which, thus far, seems not to be the case).
16. If only one site were involved, we would have no difficulty in accepting such a correlation as a satisfying explanation; and this is just the case for the original operon model of Monod and Jacob (1961). In that model, a single binding site was hypothesized which, when bound by a particular (protein) repressor, blocked the transcription of the adjacent structural genes.
17. See Yuh and Davidson (1996); Yuh, Moore, and Davidson (1996); Kirchhamer, Yuh, and Davidson (1996).
18. Yuh et al. (2001) have now extended their model to include the results of similar analyses of the structure of Module B (adjacent to Module A). Summarizing their findings, they write: "Logic considerations predicted an internal cis-regulatory switch by which spatial control of endo16 expression is shifted from Module A (early) to Module B (later). This prediction was confirmed experimentally and a distinct set of interactions in Module B that mediate the switch function was demonstrated. The endo16 computational model now provides a detailed explanation of the information processing functions executed by the cis-regulatory system of this gene throughout embryogenesis" (p. 617).
19. A similar point is made by Soraya de Chadarevian (2001) in her analysis of the role of 3-D models in research on protein structure.
20. Eric Davidson is persuaded that it cannot (conversation with Davidson, July 23, 1998). The same issue has arisen many times before—for example, in attempts to account for the evolution of complex structures such as wings or eyes, or of such complex behaviors as sex-

ual reproduction—but in none of these cases is the difficulty posed in such precise molecular terms.

21. Conversation with Eric Davidson, July 23, 1998.
22. See Hesse (1980).
23. See Chapter 9 for further discussion.
24. For one example of Odell's contributions to mathematical biology that has attracted particular notice, see Odell et al. (1981).
25. For an example of Odell's earlier work on related problems, see Edgar et al. (1987).
26. See Gilbert (1997) for a review.
27. Von Dassow et al. (2000), p. 188.
28. Ed Munro takes exception to this claim (personal communication), arguing that it is in fact possible to develop considerable intuition even for so cumbersome a system of equations.
29. Von Dassow et al. (2000), pp. 188–189.
30. Ibid.
31. Dearden and Akam (2000), pp. 131–132.
32. Von Dassow et al. (2000), p. 191.
33. See Keller (2000a), chap. 4.
34. Odell et al. (2001). Just as with the developmental module analyzed by Yuh et al. (1998; 2001), the evolution of such a network remains a considerable challenge. Computer simulations of pathways of evolution of pattern-forming networks performed by Salazar-Ciudad et al. (2001a; 2001b) suggest that stripe patterns of more than three stripes are far easier to generate by reaction-diffusion gene networks (and require far fewer genes) than by hierarchical networks of the kind seen in *Drosophila*. They therefore ask (2001b): "Why does modern-day Drosophila not use a reaction-diffusion mechanism to produce its segments?" (p. 99). Their answer is of striking interest, particularly in relation to the findings of von Dassow et al. (2000). Reaction-diffusion gene networks, they suggest, were the first to arise, but over time, selection for robustness with respect to both genetic mutations and developmental noise results in the replacement of these by hierarchical networks (2001b). "Once an optimal pattern is found," they write, "the advent of a simple hierarchic network producing part of the pattern (reinforcing one stripe against developmental or environmental noise, for example) will be immediately adaptive and will increase its frequency in the population" (p. 101). For discussion of the larger implications of this work, see Szathmáry (2001).
35. Convened in June 1992. The steering committee was made up of

mathematicians, computer scientists, and (a few) biologists, and the final report ("Mathematics and Biology: The Interface") is available on-line: http://www.bis.med.jhmi.edu/Dan/mathbio/T.html.

36. In the main body of the report (at the end of chap. 1), they add: "Our brains are incapable of coping with the wealth of biological data without the assistance of computers. The complexity of biological problems requires that we also apply mathematical and computational approaches, and the benefits of such applications will be shared equally by the disciplines of biology and mathematics."
37. "Modeling of Biological Systems," convened March 14–15, 1996, and co-chaired by Peter Kollman (UCSF) and Simon Levin (Princeton). The report can be found at: http://www.nsf.gov/bio/pubs/mobs/mobs.htm.
38. In 1998, for example, the National Science Foundation announced a major initiative aimed at "understanding and modeling complexity in biological systems"—that is, for the study of "biocomplexity." In the first phase (1999), approximately $26 million was awarded; in the second (2000), $52.5 million; and for 2001, a total of $136.3 million (representing an increase of 173 percent) has been requested in the proposed budget (totaling $4.572 billion for all divisions of the NSF). The primary focus of this initiative is on organism-environment interactions, but the announcement clearly states that the term biocomplexity is intended to be encompassing, referring "to phenomena that arise from dynamic interactions that take place within biological systems and between these systems and the physical environment" (http://www.nsf.gov/home/crssprgm/be/be_start.html). Thus, it explicitly includes investigations on the origin and dynamics of complexity in biological development; in fact, such proposals accounted for 15–20 percent of the awards during the first two years of the initiative.
39. Personal communication, January 30, 2001. Although comparable figures from the NIH were unattainable, a dramatic contrast is evident between this number and the proportion of 1983 NIH dollars devoted to mathematical models in biology (approximately 0.3 percent).
40. See Emmett (2001).
41. The number of entries listed in Medline under the keywords "mathematical" and "computational model" shows an increase of approximately 13-fold over the last 30 years. Many of these publications are directly focused on bioinformatics and the analysis of sequence data,

but in a number of new programs (Cal Tech, Cornell, and the joint effort between the University of Washington and the Fred Hutchinson Cancer Research Center), a major focus is (as in the examples discussed here) on the analysis and interpretation of the experimental data of molecular developmental genetics.

42. Of particular importance has been Stephen Wolfram's program *Mathematica* (introduced in 1988). Exploiting the power of cellular automata to simulate differential equations, Wolfram developed and marketed a user-friendly computer program that enables someone who is not literate in conventional mathematics to analyze almost any of the equations he or she would normally encounter. Soon, a number of similar programs appeared on the market (*Maple, Mathcad, Scientific Workplace,* and *Theorist*). Among mathematical biologists, programs such as *Grind* (de Boer, Utrecht) and *Biograph* (Odell) have been especially popular. What the availability of such software has achieved is the effective removal of the most obvious barrier that has historically insulated biologists from mathematics, namely, a lack of training in mathematical techniques. *Mathematica* has proven a phenomenal commercial success—Wolfram's own claim is of a million users to date (personal communication). But even if his estimate were to prove exaggerated, there is little question that programs like *Mathematica* have created a new and significantly expanded market for the use of mathematical models—in biology, in the physical and engineering sciences, and even in the history of science.
43. Dearden and Akam (2000), p. 132.
44. With a veiled allusion to Turing, John Doyle (2001) extols the achievements of the new mathematical and computational models in a commentary entitled "Beyond the Spherical Cow." Doyle urges alliances with engineering in addition to those with physics on the grounds that engineers are more like biologists than physicists "in revelling in the enormity, variety and sheer complexity of the systems they study. No interest in spherical cows here" (p. 152).
45. Galison (1997), pp. xix, 689.
46. Keller (2000e).
47. While the meaning of the term mathematics has varied extensively over the course of history, twentieth-century usage tended (especially in mathematics departments) to restrict it to the use of analytic, deductive techniques, relegating computation to the domain of applied mathematics.

48. Spengler (2000), p. 1221; von Dassow et al. (2000), p. 191.
49. Brenner (1999), p. 1964.
50. For just a few of the many other examples appearing in the recent biological literature, I might cite Jain and Krishna (2001); Palsson and Othmer (2000); Fickett and Wasserman (2000); Wagner (2000); and Oster et al. (2000). For a recent overview, see Thieffry (1999).
51. Brenner (1999), p. 1964.
52. Keller (2000a).
53. Collingwood (1940), pp. 300, 302.
54. For many workers in the field today, "use value" is often taken as the goal (and perhaps even the test) of an explanation: an explanation is expected to provide a recipe for construction; at the very least, it should provide us with effective means of intervening. Causes, in turn, are identified by their efficacy as handles. Robert Weinberg (1985), for example, suggests that the reason his colleagues are convinced that genes are causal agents of development and that "the invisible agents they study can explain . . . the complexity of life" is that, by manipulating these agents, it is now "possible to change critical elements of the biological blueprint at will" (p. 48). And in a similar vein, Phil Sharp's response to a question I put to him several years ago about the status of explanation in developmental biology is also worth noting: "We will know we have an explanation of development when we can make it happen in the lab" (private communication, 1997).
55. As Buchwald, Cartwright, Hacking, and many others have argued. Somewhat more cautiously, I would suggest that the extent to which such arguments might also help us to a better understanding of physics, particularly in the twenty-first century, remains an open question.
56. Among the many benefits researchers have begun to anticipate from such models is their use as "design tools [that] should speed the rise of a greatly heightened capability to engineer living systems"; see Endy and Brent (2001), p. 395.

9. Synthetic Biology Redux—Computer Simulation and Artificial Life

1. See also Loup Verlet's discussion of the importance to Newton of a similarly double meaning of the Latin verb *fingere* (1993), pp. 289–

290. Verlet goes on to make the more general observation: "The same slippage can be seen in such words as 'to forge' or 'to fabricate,' where perfection in imitation generates the disquiet: is it real, or is it fictional?" (p. 290, my translation). The importance of the double meaning of "to forge" is also examined in Schäffer (1997).

2. See discussion of echo simulators developed to train AI operators of an aircraft interception radar set in R. L. Garman (1942).
3. Vance (1960), pp. 1–2.
4. The use of simulations in physics is sometimes referred to as synthetic physics. Ray (1995) describes his own work on biological simulations as "An Evolutionary Approach to Synthetic Biology."
5. By now, a sizable literature has grown on simulation in the physical sciences; see Galison (1996; 1999); Rohrlich (1991); Winsberg (1999). For more extended discussions of the evolving meanings and uses of computer simulation, see Hughes (1999) and Keller (2000e).
6. Langton (1986), p. 147.
7. Langton (1989), p. 2.
8. In what was almost surely the first such use, Ted Wainwright and Berni Alder (1958) wrote: "With fast electronic computers it is possible to set up artificial many-body systems with interactions which are both simple and exactly known. Experiments with such a system can yield not only the equilibrium and transport properties at any arbitrary density and temperature of the system, but also any much more detailed information desired. With these 'controlled' experiments in simple systems it is then possible to narrow down the problem as to what analytical scheme best approximates the many-body correlations" (p. 116).
9. Hughes (2000), pp. 133–134.
10. Wolfram (1986), p. v. Wolfram's remarks at least implicitly evoke the analytic-synthetic distinction, placing traditional science in the analytic camp and his "new form of science" in the synthetic. I would suggest, however, that the distinctiveness of CA modeling would be better characterized in terms of the new orders of magnitude at which computations of the synthetic activity of many parts can now be conducted. "Emergence," that is, is better described as a property of scale than of the philosophical category synthetic per se.
11. *Lettres galantes: Oeuvres,* vol. 1, pp. 322–323, quoted in Jacob (1976), p. 63.
12. Presented at the Hixon Symposium on September 20, 1948.
13. See von Neumann (1966).
14. The Game of Life was first described by Martin Gardner in *Scientific*

American, October 1970. A fuller account of this history—one that also includes the work of Ulam, Barricelli, Holland, and many others—would certainly be welcome; George Dyson (1997) provides a good start for such an investigation.

15. For an overview of the use of CA in fluid dynamics and statistical mechanics, see Rothman and Zaleski (1997).
16. Farmer et al. (1984).
17. As Toffoli and Margolus (1987) write: "In this context, ordinary computers are of no use . . . On the other hand, the structure of a cellular automaton is ideally suited for realization on a machine having a high degree of parallelism and local and uniform interconnections" (p. 8). Conversely, however, it must also be said that the design of such machines, at least as envisioned by Hillis (1984), was itself "based on cellular automata."
18. CA advocates see exact computability as a major advantage over DEs. As Toffoli (1984) writes, "Any properties that one discovers through simulation are guaranteed to be properties of the model itself rather than a simulation artifact" (p. 120).
19. Wolfram (1986), p. v.
20. Rohrlich (1991) notes the "tendency to forget that these figures are the results of a computer simulation, of a calculation; they are not photographs of a material physical model" (p. 511). Even less are they photographs of a material biological process.
21. Use of the term simulation varies considerably in the literature, sometimes employed to denote both the representation of CA models and the models themselves; but as Hughes (2000) emphasizes, the difference is important. He marks the distinction by using simulation to refer to representations of the behavior of the CA model, and model to refer to the cellular automaton. In my terminology, however, the word representation refers to the visual display, and "simulation" describes the particular kind of model that CA computations make possible.
22. Toffoli and Margolus (1987), p. 1.
23. See, for example, Galison (1996; 1999).
24. By far the most widely distributed are the programs for simulating the evolution of virtual living universes, most famously Tom Ray's Tierra and its derivatives such as Sim-Life and Avida. Karl Sims's Sim-Life and its offshoots (Sim-City, Sim-Earth, and Sim-Ant) have been widely marketed as games that even children can play.
25. Vichniac (1994).
26. Early on, Feynman (1967) hypothesized "that ultimately physics will

not require a mathematical statement, that in the end the machinery will be revealed, and the laws will turn out to be simple, like the chequer board with all its apparent complexities" (p. 57). Today, however, Stephen Wolfram is the leading advocate of a digitally based physics. Wolfram's long-awaited magnum opus on the subject, *A New Kind of Science,* is scheduled for publication in January 2002.

27. Langton (1986), p. 147.
28. Langton (1989), pp. 1–2.
29. Langton (1989) describes these processes as "highly reminiscent of *embryological development,* in which *local hierarchies* of higher-order structures develop and *compete* with one another for support among the low-level entities" (p. xxiii, italics in original), but others such as Fleischer (1995) and Ray (1998) are using these models to explain embryological development itself.
30. Hayles (1996), pp. 146–147.
31. Ray (1995), p. 184. In fact, however, the hardware does matter, for it imposes a time scale of critical importance to human observers. Thus, hardware composed of vacuum tubes or mechanical switches would simply have been too slow for "digital organisms" to have emerged in human time.
32. Ibid., p. 184.
33. Ibid., p. 185.
34. Hayles (1996), p. 151.
35. Ray (1995), p. 185.
36. As Karl Sims (1991) explains it, "*Perceptual selection* is used because fitness functions that could determine how interesting or aesthetically pleasing a dynamical system is would be difficult to define" (p. 172).
37. Quoted in Waldrop (1992), p. 194.
38. The term neural nets refers to cellular automata models in which the strength of interactions between elements is progressively modified according to the effectiveness of the network in performing particular (pre-set) tasks. That is, connections are strengthened according to the relative success of the computations they enable.
39. Ray (1998), p. 33.
40. Lindenmayer (1968). Based on context-free rules for rewriting growth algorithms over the course of development, these models are known as L-systems.
41. Jacob (1999), p. 83.
42. First developed by Frederic Gruau (see Gruau and Whitley, 1993) and applied by J. R. Koza and his colleagues (see Koza, 1992).

43. Jacob (1999), p. 84.
44. As I argued in Chapter 1, these have mainly to do with differences in the kinds of anxieties prevailing then and now.
45. See, for example, Maynard Smith (2000). Such categorizations are difficult to make precise and may often seem simplistic, but the very fact of the recurrence (both here and in the wider literature) suggests a need for something of the sort, and perhaps especially so in this context.
46. Clark (1996).
47. See especially Oyama (1985) and Oyama et al. (2001).
48. One of the earliest arguments for the need to focus on emergent properties in understanding the distinctive features of living organisms was made by the philosopher C. D. Broad (1925).
49. Clark (1996), p. 263.
50. Rodney Brooks (1997), http://www.edge.org/documents/archive/edge31.html.
51. Grand (2000), pp. 7–8.
52. Von Neumann (1966), quoted in Pattee (1988), p. 69.
53. Interestingly, work on "directed evolution" also grew out of discussions originally held at the Santa Fe Institute. In directed evolution, enzymes designed to perform specific tasks are produced either by bacteria that have been brought into existence by sequential selection, under conditions ever more closely approximating the targeted task, or by direct selection of proteins produced by laboratory recombination of homologous genes; see, for example, Joyce (1992); (1997); Arnold (2001); and Arnold and Volkov (1999).
54. Weiss et al. (1999). This work is of particular interest because it draws its inspiration directly (and explicitly) from the early efforts of Motoyosi Sugita (1963), Stuart Kauffman (1971), and René Thomas (1973) to construct formal models of genetic regulatory networks.
55. Knight and Sussman (1998).
56. Abelson and Forbes (2000), p. 25. Dyson (1985).
57. See Mary Hesse's (1980) discussion of scientific metaphors.
58. See Moravec (1988).
59. Lange (1996), p. 231.
60. Levy (1993), p. 10.
61. In *Les mots et les choses* (1966), Foucault made the claim that, in the eighteenth century, "Life itself did not exist" (p. 139), a claim to which many historians have since objected. Joseph Schiller (1978), for example, argued, "The opposite is nearer the truth: the inanimate did not exist but life there was to excess, penetrating everywhere and

animating everything" (p. 79). I suggest, however, that Foucault's claim does make historical sense if read as a claim about "life itself," that is, as a claim about life as a natural kind.

62. Lamarck (1984 [1809]) wrote: "A study of the phenomena resulting from the existence of life in a body provides no definition of life, and shows nothing more than objects that life itself has produced. The line of study which I am about to follow has the advantage of being more exact, more direct and better fitted to illuminate the important subject under consideration; it leads, moreover, to a knowledge of the true definition of life" (p. 201).
63. Jacob (1976), p. 89.
64. It is noteworthy that, after the flurry of essays and books by that title in the early part of the twentieth century, the question "What is life?" faded from view among biologists. It was resurrected by Erwin Schroedinger with the publication of his famous book on the subject in 1943 and has remained, ever since, most commonly associated with Schroedinger's name—only rarely if ever posed by contemporary experimental biologists. To P. B. Medawar (1977), such discussions indicate "a low level in biological conversation" (p. 7). By tacit consent, today's biologists appear to concur with the judgment of Norman Pirie from the 1930s that the question is "meaningless." "Nothing turns," wrote Pirie (1938), "on whether a virus is described as a living organism or not" (p. 22). Where the question of what life *is* does arise today is mainly in A-Life studies and robotics. And like Pirie, we might ask: What hangs on whether these creatures are described as living or not, for either the scientists, the engineers, the industry, or the consumers of their products?
65. Lamarck (1984 [1809]), p. 201.
66. As Lamarck (1984 [1809]) wrote, "If we wish to arrive at a real knowledge of what constitutes life, what it consists of, what are the causes and laws which control so wonderful a natural phenomenon, and how life itself can originate those numerous and astonishing phenomena exhibited by living bodies, we must above all pay very close attention to the differences existing between inorganic and living bodies; and for this purpose a comparison must be made between the essential characters of these two kinds of bodies" (p. 191).
67. Lamarck (1984), p. 282.
68. My argument here is closely related to that of Richard Doyle (1997). Doyle claims that, instead of constituting the actual object of biology, life is (merely) its "sublime" object.
69. The reference to thunderstorms, people, and umbrellas comes from

Charles Bennett (1986): "In the modern world view, dissipation has taken over one of the functions formerly performed by God: It makes matter transcend the clod-like nature it would manifest at equilibrium, and behave instead in dramatic and unforeseen ways, molding itself for example into thunderstorms, people and umbrellas" (p. 586), while the reference to animals, armies, and vending machines is from the definition of a system in a 1950 progress report to the U.S. Air Force; see Keller (1995), pp. 90–91.

Conclusion: Understanding Development

1. Schiller, "On the Sublime."
2. See Wolpert (1994b); Rosenberg (1997).
3. Wolpert (1994b), p. 271.
4. Wray (1998), p. 1871.
5. Rosenberg (1997), p. 450.
6. See, for example, Dupré (1993); Galison and Stump (1996); Cartwright (1999a).
7. Dupré (1993), p. 7.
8. Cartwright (1999a), p. 1.

References

Abelson, Hal, and Nancy Forbes. 2000. "Amorphous Computing." *Complexity* 5 (3): 22–25.

Abraham, Tara. 2000. "'Microscopic Cybernetics': Mathematical Logic, Automata, Theory, and the Formalization of Biological Phenomena, 1936–1970." Ph.D. diss., University of Toronto.

Agutter, Paul S., P. Colm Malone, and Denys N. Wheatley. 2000. "Diffusion Theory in Biology: A Relic of Mechanistic Materialism." *Journal of the History of Biology* 33: 71–111.

Alberts, Bruce. 1998. "The Cell as a Collection of Protein Machines: Preparing the Next Generation of Molecular Biologists." *Cell* 92: 291–294.

Alexander, Jerome, ed. 1928. *Colloid Chemistry,* vol. 2. Biology and Medicine Series. New York: The Chemical Catalogue Co.

Allen, Robert Day. 1985. "New Observations on Cell Architecture and Dynamics by Video-Enhanced Contrast Optical Microscopy." *Annals of Biophysics and Biophysical Chemistry* 14: 265–290.

Allen, Robert D., Nina S. Allen, and J. L. Travis. 1981a. "Video-Enhanced Contrast, Differential Interference Contrast (AVEC-DIC) Microscopy: A New Method Capable of Analyzing Microtubule-Related Motility in the Reticulopodial Network of *Allogromia laticollaris.*" *Cell Motility* 1: 291–302.

Allen, Robert D., J. L. Travis, Nina S. Allen, and H. Yilmaz. 1981b. "Video-Enhanced Contrast Polarization (AVEC-POL) Microscopy: A New Method Applied to the Detection of Birefringence in the Motile Reticulopodial Network of *Allogromia laticollaris.*" *Cell Motility* 1: 275–289.

Apter, M. J., and Lewis Wolpert. 1965. "Cybernetics and Development." *Journal of Theoretical Biology* 8: 244–257.

Aristotle. 1993. *On Poetics.* W. Rhys Roberts, trans. *Great Books* 8: 681–699.

Arnold, Frances H. 2001. "Combinatorial and Computational Challenges for Biocatalyst Design." *Nature* 409: 253–257.

Arnold, Frances H., and Alexander A. Volkov. 1999. "Directed Evolution of Biocatalysts." *Current Opinion in Chemical Biology* 3 (1): 54–59.

Ashburner, Michael. 1993. "Epilogue" to M. Bate and A. Martinez-Arias, eds., *The Development of Drosophila melanogaster,* pp. 1493–1506. Cold Spring Harbor: Cold Spring Harbor Laboratory Press.

Aubin, D. 1998. "A Cultural History of Catastrophe and Chaos: Around the Institut des hautes études scientifiques, France." Ph.D. diss., Princeton University.

Auyang, Sunny Y. 1998. *Foundations of Complex-System Theories.* Cambridge: Cambridge University Press.

Bastian, Henry Charlton. 1870. "Facts and Reasonings Concerning the Heterogeneous Evolution of Living Things." *Nature* 2: 170–180.

Bateson, William. 1913. *Problems of Genetics.* New Haven: Yale University Press.

Bechtel, William. 1994. "Deciding on the Data: Epistemological Problems Surrounding Instruments and Research Techniques in Cell Biology." *Philosophy of Science Association 1994* 2: 167–178.

Bénard, H. 1900. "Les tourbillons cellulaires dans une nappe liquide: I. Description générale des phénomènes." *Revue générale des sciences pures et appliquées* 11: 1261–1271.

———. 1901. "Les tourbillons cellulaires dans une nappe liquide transportant de la chaleur par convection en régime permanent." *Annales de chimie et de physique,* 7th series, 23: 62–144.

Benedikt, Moriz. 1905. "Biologie générale: les origines des formes et de la vie." *Revue scientifique* 4 (14): 417–420.

Bennett, Charles H. 1986. "On the Nature and Origin of Complexity in Discrete, Homogeneous, Locally-Interacting Systems." *Foundations of Physics* 16 (6): 585–592.

Berrill, N. J. 1984. "The Pearls of Wisdom: An Exposition." *Perspectives in Biology and Medicine* 28 (1): 1–16.

Beuttner, R. 1938. *Life's Beginning on the Earth.* Baltimore: Williams & Wilkins.

Black, Max. 1962. *Models and Metaphor: Studies in Language and Philosophy.* Ithaca: Cornell University Press.

Blumich, B., and W. Kuhn, eds. 1992. *Magnetic Resonance Microscopy.* Oxford: Oxford University Press.

Bohr, Niels. 1958 [1933]. "Light and Life." In *Atomic Physics and Human Knowledge,* pp. 3–12. New York: John Wiley & Sons.

Bonner, J. T. 1996. *Sixty Years of Biology: Essays on Evolution and Development.* Princeton: Princeton University Press.

Bonner, J. T. 2002. *A Biologist's Century.* Cambridge, MA: Harvard University Press.

Bonnier, Gaston. 1907. "Sur les prétendues plantes artificielles." *Comptes Rendus de l'Académie des Sciences* 144: 55–58.

Bookstein, Fred L. 1997. *Morphometric Tools for Landmark Data: Geometry and Biology.* Cambridge: Cambridge University Press.

Brandon, Robert N. 1990. *Adaptation and Environment.* Princeton: Princeton University Press.

Brenner, Sydney. 1999. "Theoretical Biology in the Third Millennium." *Philosophical Transactions of the Royal Society, London, B* 354: 1963–1965.

Brink, R. A. 1927. "Genetics and the Problems of Development." *American Naturalist* 61 (574): 280–283.

Broad, C. D. 1925. *The Mind and Its Place in Nature.* London: Routledge & K. Paul.

Buchwald, Jed Z., ed. 1995. *Scientific Practice: Theories and Stories of Doing Physics.* Chicago: University of Chicago Press.

Butcher, W. Deane. 1911. "Preface" to Leduc 1911, pp. i–ii.

Carr, Kimberly. 1994. "Life after Transcription." *Nature* 369: 440–441.

Cartwright, Nancy. 1997. "Models: Blueprints for Laws." *Philosophy of Science* 64: S29–2-S303.

———. 1999a. *The Dappled World: A Study of the Boundaries of Science.* Cambridge: Cambridge University Press.

———. 1999b. "Models and the Limits of Theory." In Morgan and Morrison 1999, pp. 241–281.

Cartwright, Nancy, Towfic Shomar, and Maricio Suárez. 1995. "The Tool Box of Science." *Poznan Studies in the Philosophy of the Sciences and the Humanities* 44: 137–149.

Castets, V., E. Dulos, J. Boissonade, and P. De Kepper. 1990. "Experimental Evidence of a Sustained Standing Turing-Type Nonequilibrium Chemical Pattern." *Physical Review Letters* 64: 2953.

Cat, Jordi. 2000. "On Understanding: Maxwell on the Methods of Illustration and Scientific Metaphor." *Studies in the History and Philosophy of Modern Physics* 32 (3): 395–441.

Cetina, Karen Knorr. 1999. *Epistemic Cultures.* Cambridge: Harvard University Press.

Chambers, Robert, and Edward L. Chambers. 1961. *Explorations into the Nature of the Living Cell.* Cambridge: Harvard University Press.

Charrin and Goupil. 1907. "Absence de nutrition dans la formation des plantes artificielles de Leduc." *Comptes Rendus de l'Académie des Sciences* 144: 136–138.

Clark, Andy. 1996. "Happy Couplings: Emergence and Explanatory Interlock." In Margaret A. Boden, ed., *The Philosophy of Artificial Life,* pp. 262–281. Oxford: Oxford University Press.

Clarke, Adele, and Joan Fujimura, eds. 1992. *The Right Tools for the Job: At Work in Twentieth-Century Life Sciences.* Princeton: Princeton University Press.

Cohn, Melvin. 1956. "Added Comment." In Gaebler 1956, pp. 41–46.

———. 1958. "On the Differentiation of a Population of *Escherichia Coli* with Respect to β-galactosidase Formation." In Gaebler 1956, pp. 458–468.

Collingwood, R. C. 1940. "Causation in Practical Natural Science." In R. C. Collingwood, *An Essay on Metaphysics,* pp. 296–312. Oxford: Clarendon Press.

Conrad, Michael. 1996. "Childhood, Boyhood, Youth." *Society of Mathematical Biology Newsletter* 9 (3): 8–9.

Cornell-Bell, A. H., S. M. Finkbeiner, M. S. Cooper, and S. J. Smith. 1990. "Glutamate Induces Calcium Waves in Cultured Astrocytes: Long-Range Glial Signaling." *Science* 247 (4941): 470–473.

Creager, Angela N. H., and Jean-Paul Gaudillière. 1996. "Meanings in Search of Experiments or *Vice-versa:* The Invention of *Allosteric Regulation* in Paris and Berkeley." *History Studies in the Physical and Biological Sciences* 27 (1): 1–90.

Creath, Richard, and Jane Maienschein. 1998. *Biology and Epistemology.* New York: Cambridge University Press.

Crick, Francis. 1957. "On Protein Synthesis." *Symposia of the Society for Experimental Biology* 12: 138–163.

Crombie, A. C. 1994. *Styles of Scientific Thinking in the European Tradition: The History of Argument and Explanation Especially in the Mathematical and Biomedical Sciences and Arts.* London: Duckworth.

Cuénot, Lucien. 1911. *Revue générale des sciences* 12: 41–42.

———. 1913. *Revue générale des sciences* 24: 556.

Daston, Lorraine. 1991. "Marvelous Facts and Miraculous Evidence in Early Modern Europe." In J. Chandler, A. I. Davidson, and H. Harootunian, eds., *Questions of Evidence: Proof, Practice, and Persuasion across the Disciplines,* pp. 243–274. Chicago: University of Chicago Press.

Davidson, Eric H. 1986. *Gene Activity in Early Development.* 1st ed. 1968; 2nd 1976. Orlando: Academic Press.

Davis, Bernard D. 1961. "Opening Address: The Teleonomic Significance of Biosynthetic Control Mechanisms." *Cold Spring Harbor Symposia on Quantitative Biology* 26: 1–10.

De Chadarevian, Soraya. 2001. "Models and the Making of Molecular Biology." In S. de Chadarevian and N. Hopwood, eds., *Displaying the Third Dimension: Models in the Sciences, Technology, and Medicine.* Stanford: Stanford University Press.

De Man, Paul. 1978. "The Epistemology of Metaphor." *Critical Inquiry* 5 (1): 13–30.

Dean, Bashford. 1911. Review of *Théorie physico-chimique de la vie et générations spontanées. Science* 33: 304–305.

Dearden, Peter, and Michael Akam. 2000. "Segmentation *in silico.*" *Nature* 406: 131–132.

Delbrück, Max. 1949a. "A Physicist Looks at Biology." Rpt. in Cairns et al., *Phage and the Origins of Molecular Biology,* pp. 9–22. Cold Spring Harbor: Cold Spring Harbor Press, 1966.

———. 1949b. "Discussion." In Max Delbrück, trans., *Unités biologiques douées de continuité génétique,* p. 33. Lyon: Éditions du CNRS. [Ms. sent to Leo Szilard, February 1, 1960.]

Denbigh, K. G., Margaret Hicks, and F. M. Page. 1948. "The Kinetics of Open Reaction Systems." *Transactions of the Faraday Society* 44: 479–494.

Derrida, Jacques. 1982 [1971]. "White Mythology: Metaphor in the Text of Philosophy." In Alan Bass, trans., *Margins of Philosophy,* pp. 207–271. Chicago: University of Chicago Press; originally published in *Poétique* 5 (1971).

Dobell, Clifford. "D'Arcy Wentworth Thompson." *Obituary Notices of the Royal Society of London* 1949: 612n18.

Doyle, John. 2001. "Computational Biology: Beyond the Spherical Cow." *Nature* 411 (6834): 151–152.

Doyle, Richard. 1997. *On Beyond Living.* Stanford: Stanford University Press.

Driever, W., and C. Nüsslein-Volhard. 1988a. "A Gradient of Bicoid Protein in *Drosophila* Embryos." *Cell* 54: 83–93.

———. 1988b. "The Bicoid Protein Determines Position in the *Drosophila* Embryo in a Concentration-Dependent Manner." *Cell* 54: 95–104.

Dulos, E., J. Boissonade, J. J. Perraud, B. Rudovics, and P. De Kepper. 1996. "Chemical Morphogenesis: Turing Patterns in an Experimental Chemical System." *Acta Biotheoretica* 44 (3–4): 249–261.

Dupré, John. 1993. *The Disorder of Things: Metaphysical Foundations of the Disunity of Science.* Cambridge: Harvard University Press.

Dyson, Freeman. 1985. *Infinite in All Directions.* New York: Harper & Row.

Dyson, George. 1997. *Darwin among the Machines: The Evolution of Global Intelligence.* New York: Addison Wesley.

Easlea, Brian. 1981. *Science and Sexual Oppression.* London: Weidenfeld and Nicolson.

Ebert, James D., and Fred H. Wilt. 1960. "Animal Viruses and Embryos." *Quarterly Review of Biology* 35: 261–312.

Edgar, B. A., G. M. Odell, and G. Schubiger. 1987. "Cytoarchitecture and the Patterning of *Fushi tarazu* Expression in the *Drosophila* Blastoderm." *Genes & Development* 10: 1226–1237.

Edwards, Paul N. 1996. *The Closed World: Computers and the Politics of Discourse in Cold War America.* Cambridge: MIT Press.

Emmett, Arielle. 2001. "Wanted: Applicants for NIGMS Grants: Lack of Computational Brainpower Presents Enigma." *The Scientist* 15 (1): 1–5.

Endy, Drew, and Roger Brent. 2001. "Modelling Cellular Behaviour." *Nature* 409: 391–395.

Engel, Andreas, Yuri Lyubchenko, and Daniel Miller. 1999. "Atomic Force Microscopy: A Powerful Tool to Observe Biomolecules at Work." *Trends in Cell Biology* 9: 77–80.

Ephrussi, Boris. 1956. "Enzymes in Cellular Differentiation." In Gaebler 1956, pp. 29–40.

Farley, John. 1974. *The Spontaneous Generation Controversy from Descartes to Oparin.* Baltimore: Johns Hopkins Press.

Farmer, Doyne, Tommaso Toffoli, and Stephen Wolfram. 1984. *Cellular Automata: Proceedings of an Interdisciplinary Workshop, Los Alamos March 7–11, 1983.* Amsterdam: North-Holland.

Feininger, Andreas. 1979 [1956]. *The Anatomy of Nature.* New York: Dover.

Feynman, Richard. 1967. *The Character of Physical Law.* Cambridge: MIT Press.

Fickett, J. W., and W. W. Wasserman. 2000. "Discovery and Modeling of Transcriptional Regulatory Regions." *Current Opinion in Biotechnology* 11 (1): 19–24.

Fine, Arthur. 1993. "Fictionalism." *Midwest Studies in Philosophy* 18: 1–18.

Fleischer, Kurt W. 1995. "A Multiple-Mechanism Developmental Model for Defining Self-Organizing Geometric Structures." Ph.D. diss., California Institute of Technology.

Foe, Victoria E., Christine M. Field, and Garrett M. Odell. 2000.

"Microtubules and Mitotic Cycle Phase Modulate Spatiotemporal Distributions of F-actin and Myosin II in Drosophila Syncytial Blastoderm Embryos." *Development* 127 (9): 1767–1787.

Foucault, Michel. 1966. *Les mots et les choses: une archéologie des sciences humaines*. Paris: Gallimard.

Fraser, S. E., and R. M. Harland. 2000. "The Molecular Metamorphosis of Experimental Embryology." *Cell* 100 (1): 41–55.

French, V., P. J. Bryant, and S. V. Bryant. 1976. "Pattern Regulation in Epimorphic Fields." *Science* 193: 969–981.

Freud, Sigmund. 1950. *Collected Papers of Sigmund Freud, 1856–1939.* James Strachey, trans. London: Hogarth Press and Institute of Psychoanalysis.

Gadamer, Hans-Georg. 1975. *Truth and Method.* G. Barden and J. Cumming, trans. London: Sheed & Ward.

Gaebler, Oliver H. 1956. *Enzymes: Units of Biological Structure and Function.* New York: Academic Press.

Galison, Peter. 1994. "The Ontology of the Enemy: Norbert Wiener and the Cybernetic Vision." *Critical Inquiry* 21: 228–266.

———. 1996. "Computer Simulations and the Trading Zone." In P. Galison and D. J Stump, eds., *The Disunity of Science: Boundaries, Contexts, and Power,* pp. 118–157. Stanford: Stanford University Press.

———. 1997. *Image and Logic: A Material Culture of Microphysics.* Chicago: University of Chicago Press.

Galison, Peter, and David J. Stump, eds. 1996. *The Disunity of Science: Boundaries, Contexts, and Power.* Stanford: Stanford University Press.

Garman, R. L. 1942. "AI-10 Trainer Simulation at I. F. Level." Report 105-1, Radiation Laboratory, MIT, August 15, 1942.

Ghyka, M. 1946. *The Geometry of Art and Life.* New York: Sheed and Ward.

Gierer, A., and H. Meinhardt. 1972. "A Theory of Biological Pattern Formation." *Kybernetik* 12 (1): 30–39.

Gilbert, Scott F. 1997. *Developmental Biology.* 5th ed. Sunderland, MA: Sinauer Associates.

Goethe, Johann Wolfgang von. 1983. "The Enterprise Justified." In *On Morphology,* trans. Douglas Miller. *Scientific Studies,* vol. 12, pp. 61–62. New York: Suhrkamp.

Goodwin, B. C., and S. A. Kauffman. 1990. "Spatial Harmonics and Pattern Specification in Early *Drosophila* Development. Part I. Bifurcation Sequences and Gene Expression." *Journal of Theoretical Biology* 144: 303–319.

Gould, Stephen Jay. 1976. "D'Arcy Thompson and the Science of Form." In M. Grene and E. Mendelsohn, eds., *Topics in the Philosophy of Biology,* pp. 66–97. Dordrecht: D. Reidel.

Goupil, Michelle. 1991. *Du flou au clair? L'histoire de l'affinité chimique.* Paris: Editions du CTHS.

Gradenwitz, Alfred. 1907. "Leduc's Artificial Plants and Cells." *Scientific American* 96: 234–236.

———, trans. 1912. *Das leben in seinem physikalisch-chemischen zusammenhang* / von dr. Stéphane Leduc. Halle: L. Hofstetter.

Grand, Steve. 2000. *Creation: Life and How to Make It.* London: Weidenfeld and Nicolson.

Gruau, F., and D. Whitley. 1993. "Adding Learning to the Cellular Development of Neural Networks: Evolution and the Baldwin Effect." *Evolutionary Computation* 1 (3): 213–233.

Gruenberg, Benjamin C. 1911a. "The Creation of 'Artificial Life.'" *Scientific American* 105: 231–237.

———. 1911b. "Artificial Life II: Making the Non-Living Do the Work of the Living." *Scientific American* 105: 272–286.

Hacking, Ian. 1983. *Representing and Intervening.* Cambridge: Cambridge University Press.

———. 1987. "Weapons Research and the Form of Scientific Knowledge." *Canadian Journal of Philosophy* (supplement) 12: 237–262.

———. 1992. "'Style' for Historians and Philosophers." *Studies in the History and Philosophy of Science* 23 (1): 1–20.

Haeckel, Ernst. 1892. *The History of Creation,* vol. 2. New York: D. Appleton.

Haldane, J. B. S. 1940. "Unsolved Problems of Science: Can We Make Life?" In *Keeping Cool and Other Essays,* pp. 23–30. London: Chatto & Windus.

Hartman, Philip E., and Sigmund R. Suskind. 1965. *Gene Action.* Englewood Cliffs, NJ: Prentice-Hall.

Hayles, N. Katherine. 1996. "Narratives of Artificial Life." In George Robertson, Melinda Mash, Lisa Tickner, Jon Bird, Barry Curtis, and Tim Putnam, eds., *FutureNatural: Nature, Science, Culture,* pp. 146–164. London: Routledge.

Hearst, J. E. 1990. "Microscopy: 'Seeing Is Believing.'" *Nature* 347 (6290): 230.

Heidegger, Martin. 1977. "The Age of the World Picture." In William Lovitt, trans., *The Question Concerning Technology and Other Essays,* pp. 115–154. New York: Harper and Row.

Herrera, Alphonso L. 1928. "Plasmogeny." In Alexander 1928, pp. 81–91.

Herschman, H. R., D. C. MacLaren, M. Iyer, et al. 2000. "Seeing Is Believing: Non-invasive, Quantitative and Repetitive Imaging of Reporter Gene Expression in Living Animals, Using Positron Emission Tomography." *Journal of Neuroscience Research* 59 (6): 699–705.

Hesse, Mary. 1966. *Models and Analogies in Science.* Notre Dame: University of Notre Dame Press.

———. 1980. "The Explanatory Function of Metaphor." In Mary Hesse, *Revolutions and Reconstructions in the Philosophy of Science,* pp. 111–124. Bloomington: Indiana University Press.

———. 1987. "Tropical Talk: The Myth of the Literal." *The Aristotelian Society* 61 (supplement): 297–310.

———. 1993. "Models, Metaphors and Truth." In F. R. Ankersmit and J. J. A. Mooij, eds., *Knowledge and Language,* vol. 3. Dordrecht: Kluwer Academic Publishers.

Hillis, Daniel W. 1984. "The Connection Machine: A Computer Architecture Based on Cellular Automata." *Physica* 10D: 213–218.

———. 1993. "Why Physicists Like Models, and Why Biologists Should." *Current Biology* 3 (2): 79–81.

Hodges, Alan. 1983. *Alan Turing: The Enigma of Intelligence.* London: Burnett Books.

Hoke, Franklin. 1993. "Confocal Microscopy: Viewing Cells as 'Wild Animals.'" *The Scientist* 7 (2): 17–19.

Holmes, Frederic L. 1995. "The Life Cycles: The Regulation of Intermediary Metabolism." Unpub. ms. presented at the 1995 meetings of the International Society for History, Philosophy, and Social Studies of Biology, Leuven, Belgium.

Hooke, Robert. 1665. *Micrographia; or, Some Physiological Descriptions of Minute Bodies Made by Magnifying Glasses.* London: J. Martyn and J. Allestry.

Horder, T. J., J. A. Witkowsky, and C. C. Wylie, eds. 1986. *A History of Embryology.* Cambridge: Cambridge University Press.

Horowitz, N., and H. K. Mitchell. 1950. *Annual Review of Biochemistry* 20: 465–486.

Hughes, R. I. G. 1999. "The Ising Model, Computer Simulation, and Universal Physics." In Morgan and Morrison 1999, pp. 97–145.

Humphries, Sally. 1993. "From Riddle to Rigour: Satisfactions of Scientific Prose in Ancient Greece." In S. Marchand and E. Lunbeck, eds., *Proof and Persuasion: Essays on Authority, Objectivity, and Evidence,* pp. 3–24. Turnhout: Brepols.

Huta, Carole. 1998. "Jean Sénébier (1742–1809): Un dialogue entre l'ombre et la lumière: L'art d'observer à la fin du XVIIIe siècle." *Revue d'histoire des sciences et de leurs applications* 51: 93–105.

Hutchison, G. Evelyn. 1948. "In Memoriam, D'Arcy Wentworth Thompson." *American Scientist* 36: 577–606.

Huxley, Thomas H. 1870. "Biogenesis and Abiogenesis." Rpt. in T. H. Huxley, *Discourses: Biological and Geological Essays.* New York: Appleton, 1894.

Inoué, Shinya. 1953. "Polarization Optical Studies of the Mitotic Spindle." *Chromosoma* (supplement) 5: 487–500.

———. 1981. "Cell Division and the Mitotic Spindle." *Journal of Cell Biology* 89: 131s–147s.

———. 1993. "Keith Porter and the Fine Architecture of Living Cells." In Robert B. Barlow, Jr., John E. Dowling, and Gerald Weissmann, eds., *The Biological Century: Friday Evening Talks at the Marine Biological Laboratory,* pp. 100–114. Woods Hole, MA: Marine Biological Laboratory.

Inoué, Shinya, and K. R. Spring. 1997. *Video Microscopy.* New York: Plenum Press.

Israel, Giorgio. 1993. "The Emergence of Biomathematics and the Case of Population Dynamics." *Science in Context* 6: 469–509.

Jacob, Christian. 1999. "The Art of Genetic Programming. *IEEE Intelligent Systems,* May/June: 83–84.

Jacob, François. 1976 [1970]. *The Logic of Life*. New York: Pantheon.

Jacob, F., and J. Monod. 1959. "Gènes de structure et gènes de régulation dans la biosynthèse des protèines." *Comptes Rendus de l'Académie des Sciences* 349: 1282–1284.

———. 1961. "Genetic Regulatory Mechanisms in the Synthesis of Proteins." *Journal of Molecular Biology* 3: 318–356.

———. 1963. "Genetic Repression, Allosteric Inhibition, and Cellular Differentiation." In Michael Locke, ed., *Cytodifferentiation and Macromolecular Synthesis,* pp. 30–64. New York: Academic Press.

Jacobs, Russell E., Eric T. Ahrens, Thomas J. Meade, and Scott E. Fraser. 1999. "Looking Deeper into Vertebrate Development." *Trends in Cell Biology* 9: 73–76.

Jain, Sanjay, and Sandeep Krishna. 2001. "A Model for the Emergence of Cooperation: Interdependence and Structure in Evolving Networks." *Proceedings of the National Academy of Sciences USA* 98: 543–547.

Jay, Martin. 1988. "Scopic Regimes of Modernity." In Hal Foster, ed., *Vision and Visuality,* pp. 3–27. Seattle: Bay Press.

Jentsch, Ernst. 1906. "Zur Psychologie des Unheimlichen." *Psychiatrisch-Neurologiches Wochenschrift* 8: 195–198, 203–205.

Joyce, Gerald F. 1992. "Directed Molecular Evolution." *Scientific American* 267 (6) (December): 48–55.

———. 1997. "Evolutionary Chemistry: Getting There from Here." *Science* 276: 1658–1659.

Judson, Horace. 1979. *The Eighth Day of Creation.* New York: Simon and Schuster.

Kant, Immanuel. 1786. *Metaphysical Foundations of Natural Science.* Trans. James Ellington. Indianapolis: Bobbs-Merrill, 1970.

Kauffman, S. A. 1969. "Homeostasis and Differentiation in Random Genetic Control Networks." *Nature* 224 (215): 177–178.

———. 1971. "Gene Regulation Networks: A Theory for Their Global Structure and Behavior." *Current Topics in Developmental Biology* 6: 145–182.

———. 1973. "Control Circuits for Determination and Transdetermination." *Science* 181 (97): 310–318.

———. 1981. "Pattern Formation in the Drosophila Embryo." *Philosophical Transactions of the Royal Society, London, B* 295: 567–594.

Kay, Lily. 2000. *Who Wrote the Book of Life? A History of the Genetic Code*. Stanford: Stanford University Press.

Keller, E. F. 1983. "The Force of the Pacemaker Concept in Theories of Aggregation in Cellular Slime Mold." *Perspectives in Biology and Medicine* 26 (4): 515–521.

———. 1985. *Reflections on Gender and Science*. New Haven: Yale University Press.

———. 1986. "Making Gender Visible in the Pursuit of Nature's Secrets." In T. de Laurentis, ed., *Feminist Studies/Critical Studies*, pp. 67–77. Bloomington: University of Indiana Press.

———. 1992. "Critical Silences in Scientific Discourse: Problems of Form and Re-Form." In *Secrets of Life, Secrets of Death*, pp. 73–92. New York: Routledge.

———. 1994. "Rethinking the Meaning of Genetic Determinism." *Tanner Lectures on Human Values* 15: 113–139.

———. 1995. *Refiguring Life: Metaphors of Twentieth Century Biology*. New York: Columbia University Press.

———. 1996a. "*Drosophila* Embryos as Transitional Objects: The Work of Donald Poulson and Christiane Nüsslein-Volhard." *History Studies in the Physical and Biological Sciences* 26 (2): 313–346.

———. 1996b. "The Biological Gaze." In George Robertson, Melinda Mash, Lisa Tickner, Jon Bird, Barry Curtis, and Tim Putnam, eds., *FutureNatural: Nature, Science, Culture*, pp. 107–121. London: Routledge.

———. 1998. "Making Sense of Life: Explanation in Developmental Biology." In Creath and Maienschein 1998, pp. 244–260.

———. 1999. "Understanding Development." *Biology and Philosophy* 14: 321–330.

———. 2000a. *The Century of the Gene*. Cambridge: Harvard University Press.

———. 2000b. "Is There an Organism in This Text?" In P. R. Sloan, ed., *Controlling Our Destinies: Historical, Philosophical, Ethical, and Theological Perspectives on the Human Genome Project*, pp. 273–290. Notre Dame: University of Notre Dame Press.

———. 2000c. "Models Of and Models For: Theory and Practice in Contemporary Biology." *Philosophy of Science* 67 (Proceedings): S72–S86.

———. 2000d. "Linking Organisms and Computers: Theory and Practice in Contemporary Biology." In Margery Arent Safir, ed., *Connecting Creations*. Santiago de Compostela: Centro Galego de Arte Contemporánea.

———. 2000e. "Models, Simulation, and 'Computer Experiments.'" Presented at a Workshop on the Philosophy of Scientific Experimentation, June 15–17, Amsterdam.

———. 2002. "Marrying the Pre-Modern to the Post-Modern: Computers and Organisms after WWII." In Darren Tofts, Annemarie Jonson, and Alessio Cavallaro, eds., *Prefiguring Cyberculture*. Sydney: Power Publications.

Keller, E. F., and C. Grontkowski. 1983. "The Mind's Eye." In S. Harding and M. Hintikka, eds., *Discovering Reality: Feminist Perspectives in Epistemology, Methodology and Metaphysics,* pp. 207–224. Dordrecht: D. Reidel.

Keller, E. F., and Lee A. Segel. 1970. "Slime Mold Aggregation Viewed as an Instability." *Journal of Theoretical Biology* 26: 399–415.

Kemp, Martin. 1995. "Spirals of Life: D'Arcy Thompson and Theodore Cook, with Leonardo and Dürer in Retrospect." *Physics* 32: 37–54.

Kingsland, Sharon E. 1985. *Modeling Nature: Episodes in the History of Population Biology.* Chicago: University of Chicago Press.

Kirchhamer, C. V., C. H. Yuh, and E. H. Davidson. 1996. "Modular cis-Regulatory Organization of Developmentally Expressed Genes: Two Genes Transcribed Territorially in the Sea Urchin Embryo, and Additional Examples." *Proceedings of the National Academy of Sciences USA* 93: 9322–9328.

Kirschner, Marc, John Gerhart, and Tim Mitchison. 2000. "Molecular 'Vitalism.'" *Cell* 100: 79–88.

Kitcher, Philip, and Wesley Salmon, eds. 1989. *Scientific Explanation.* Minneapolis: University of Minnesota Press.

Kitching, John A., ed. 1954. *Recent Developments in Cell Physiology: Proceedings of the Seventh Symposium of the Colston Research Society, University of Bristol.* New York: Academic Press.

Knight, Thomas F., Jr., and Gerald Jay Sussman. 1998. "Cellular Gate Technology." In Cristian Calude, J. L. Casti, and M. J. Dinneen, eds., *Unconventional Models of Computation,* pp. 257–272. New York: Springer.

Kohler, R. E. 1994. *Lords of the Fly: Drosophila Genetics and the Experimental Life.* Chicago: University of Chicago Press.

Kolmagoroff, A., I. Petrovski, and N. Piskunov. 1937. "Étude de l'équation de la diffusion avec croissance de la quantité de matière et son application à un problème biologique." *Bulletin de l'Université d'État à Moscou, série international* 1 (A): 1–25.

Kondo, Shigeru, and Rihito Asai. 1995. "A Reaction-Diffusion Wave on the Skin of the Marine Angelfish *Pomacanthus.*" *Nature* 376: 765–768.

Kopell, Nancy, and Lou Howard. 1973. "Horizontal Bands in the Belousov Reaction." *Science* 180: 1171–1173.

Kostitzin, V. A. 1937. *Biologie mathématique.* Paris: Colim.

Koza, J. R. 1992. *Genetic Programming: On the Programming of Computers by Natural Selection.* Cambridge: MIT Press.

Krebs, Hans. 1957. "Steering of Metabolic Processes." *Endeavour* 16 (63): 125–132.

Krieger, Martin H. 1992. *Doing Physics: How Physicists Take Hold of the World.* Bloomington: Indiana University Press.

Kulesa, P. M., M. Bronner-Fraser, and S. E. Fraser. 2000. "In Ovo Time-Lapse Analysis after Dorsal Neural Tube Ablation Shows Rerouting of Chick Hindbrain Neural Crest." *Development* 127 (13): 2843–2852.

Kulesa, P. M., and S. E. Fraser. 1998. "Neural Crest Cell Dynamics Revealed by Time-Lapse Video Microscopy of Whole Embryo Chick Explant Cultures." *Developmental Biology* 204 (2): 327–344.

———. 1999. "Confocal Imaging of Living Cells in Intact Embryos." *Methods in Molecular Biology* 122: 205–222.

———. 2000. "In Ovo Time-Lapse Analysis of Chick Hindbrain Neural Crest Cell Migration Shows Cell Interactions during Migration to the Branchial Arches." *Development* 127 (6): 1161–1172.

Lacalli, T. C. 1990. "Modeling the Drosophila Pair-Rule Pattern by Reaction-Diffusion: Gap Input and Pattern Control in a 4-Morphogen System." *Journal of Theoretical Biology* 144 (2): 171–194.

Lacalli, T. C., D. A. Wilkinson, and L. G. Harrison. 1988. "Theoretical Aspects of Stripe Formation in Relation to Drosophila Segmentation." *Development* 104 (1): 105–113.

Lamarck, J.-B. 1984 [1809]. *Philosophical Zoology: An Exposition with Regard to the Natural History of Animals.* Chicago: University of Chicago Press.

Landecker, Hannah. 1999. "Technologies of Living Substance: Tissue Culture in Twentieth Century Biomedicine." Ph.D. diss., Massachusetts Institute of Technology.

Lange, Marc. 1996. "Life, 'Artificial Life' and Scientific Explanation." *Philosophy of Science* 63: 225–244.

Langton, C. G. 1984. "Self-Reproduction in Cellular Automata." *Physica* 10D: 135–144.

———. 1986. "Studying Artificial Life with Cellular Automata." *Physica* 22D: 120–149.

Langton, C. G., ed. 1989. *Santa Fe Studies in the Sciences of Complexity,* vol. 6: *Artificial Life.* Reading, MA: Addison-Wesley.

Laszlo, Pierre. 1993. *La Parole des choses, ou, Le Langage de la chimie.* Paris: Hermann.

Lawrence, Peter A. 1992. *The Making of a Fly: The Genetics of Animal Design.* Oxford: Blackwell Science.

Leduc, Stéphane. 1904. "Diffusion des liquides: son rôle biologique." *Comptes Rendus de l'Académie des Sciences* 139: 986.

———. 1905. "Germination et croissance de la cellule artificielle." *Comptes Rendus de l'Académie des Sciences* 141: 2801.

———. 1906a. "Culture de la cellule artificielle." *Comptes Rendus de l'Académie des Sciences* 143: 842–844.

———. 1906b. *Les bases physiques de la vie et la Biogenèse.* Paris: Masson.

———. 1907. "Croissances artificielles." *Comptes Rendus de l'Académie des Sciences* 144: 39–41.

———. 1909. *Les croissances osmotiques et l'origine des êtres vivants.* Bar-le-duc: Jolibois.

———. 1910. *Théorie physico-chimique de la vie et générations spontanées.* Paris: A. Poinat.

———. 1911. *The Mechanism of Life.* W. D. Butcher, trans. New York: Rebman; London: Heinemann.

———. 1912. *Études de biophysique: La Biologie synthétique, étude de biophysique*. Paris: A. Poinat.

———. 1928. "Solutions and Life." In J. Alexander, ed., *Colloid Chemistry: Theoretical and Applied*. New York: The Chemical Catalogue Co.

Levin, David Michael, ed. 1993. *Modernity and the Hegemony of Vision*. Berkeley: University of California Press.

Levy, Steven. 1993. *Artificial Life: The Quest for a New Creation*. Nw York: Pantheon Books.

Lewis, E. B. 1976. "A Gene Complex Controlling Segmentation in *Drosophila*." *Nature* 276: 565–570.

Lillie, Ralph S. 1917. "The Formation of Structures Resembling Organic Growths by Means of Electrolytic Local Action in Metals, and the General Physiological Significance and Control of This Type of Action." *Biological Bulletin* 33: 135–186.

———. 1922. "Growth in Living and Non-living Systems." *Scientific Monthly* 14 (8): 113–130.

Lillie, Ralph S., and Earl N. Johnston. 1919. "Precipitation-Structures Simulating Organic Growth II: A Contribution to the Physico-Chemical Analysis of Growth and Heredity." *Biological Bulletin* 36: 225–260.

Lindenmayer, A. 1968. "Mathematical Models for Cellular Interaction in Development, Parts I and II." *Journal of Theoretical Biology* 18: 280–315.

Locke, John. 1993 [1689]. *An Essay Concerning Human Understanding*. John W. Yolton, ed. Rutland, VT: Dent.

Loeb, Jacques. 1964 [1912]. *The Mechanistic Conception of Life*. Cambridge: Harvard University Press.

Lotka, A. J. 1934. "Théorie analytique des associations biologiques." Paris: Hermann.

Lwoff, André. 1962. *Biological Order*. Cambridge: MIT Press.

Lynch, Michael, and Steve Woolgar, eds. 1990. *Representation in Scientific Practice*. Cambridge: MIT Press.

MacColl, Le Roy Archibald. 1945. *Fundamental Theory of Servomechanisms*. New York: D. Van Nostrand Company.

Macfarlane, John Muirhead. 1918. *The Causes and Course of Organic Evolution*. New York: Macmillan.

Maienschein, Jane. 1997. "Changing Conceptions of Organization and Induction." In *Forces in Developmental Biology Research: Then and Now. American Zoologist* 37 (special issue): 213–310.

Maini, Philip K., Kevin J. Painter, and Helene Nguyen Phong Chau. 1997. "Spatial Pattern Formation in Chemical and Biological Systems." *Journal of the Chemical Society, Faraday Transactions* 93 (20): 3601–3610.

Mann, Thomas. 1948. *Doctor Faustus: The Life of the German Composer Adrian Leverkühn, as Told by a Friend.* New York: A. A. Knopf.

Matus, Andrew. 1999. "Introduction: GFP Illuminates Everything." *Trends in Cell Biology* 9: 43.

Maynard Smith, John. 1998. *Shaping Life—Genes, Embryos and Evolution.* London: Weidenfeld & Nicolson.

———. 2000. "The Cheshire Cat's DNA." *New York Review of Books,* December 21, p. 43.

Maynard Smith, J., and K. C. Sondhi. 1961. "The Arrangement of Bristles in *Drosophila.*" *Journal of Embryology and Experimental Morphology* 9: 661–672.

Mayr, Ernst. 1961. "Cause and Effect in Biology." *Science* 134: 1501–1506.

McClay, David. 2000. "The Role of Thin Filopodia in Motility and Morphogenesis." *Experimental Cell Research* 253 (2): 296–301.

McGhee, George R. 1999. *Theoretical Morphology: The Concept and Its Applications.* New York: Columbia University Press.

Medawar, P. B. 1958. "D'Arcy Thompson and '*Growth and Form.*'" Postscript to Thompson 1958, pp. 219–233.

———. 1977. *The Life Science: Current Ideas of Biology.* New York: Harper & Row.

Mehta, Amit D., Matthias Rief, James A. Spudich, David A. Smith, and Robert M. Simmons. 1999. "Single-Molecule Biomechanics with Optical Methods." *Science* 283 (5408): 1689–1695.

Meinhardt, Hans. 1977. "A Model of Pattern Formation in Insect Embryogenesis." *Journal of Cell Science* 23: 117–139.

———. 1982. *Models of Biological Pattern Formation.* New York: Academic Press.

———. 1995. "Dynamics of Stripe Formation." *Nature* 376: 722–723.

Meinhardt, Hans, and Alfred Gierer. 2000. "Pattern Formation by Local Self-Activation and Lateral Inhibition." *BioEssays* 22: 753–760.

Miller, J., S. E. Fraser, and D. McClay. 1995. "Dynamics of Thin Filopodia during Sea Urchin Gastrulation." *Development* 121 (8): 2501–2511.

Mitchell, P. Chalmers. 1911. "Abiogenesis." *Encyclopedia Britannica* 1: 64–65.

Mitchison, G. J., and M. Wilcox. 1973. "Alteration in Heterocyst Pattern of *Anabaena* Produced by 7-azatryptophan." *Nature New Biology* 246 (155): 229–233.

Mitman, Gregory. 1999. *Reel Nature: America's Romance with Wildlife on Film*. Cambridge: Harvard University Press.

Mohler, William A. 1999. "Visual Reality: Using Computer Reconstruction and Animation to Magnify the Microscopist's Perception." *Molecular Biology of the Cell* 10: 3061–3065.

Monod, J., and F. Jacob. 1961. "General Conclusions: Teleonomic Mechanisms in Cellular Metabolism, Growth, and Differentiation." *Cold Spring Harbor Symposia on Quantitative Biology* 26: 389–401.

Monteith, G. R. 2000. "Seeing Is Believing: Recent Trends in the Measurement of Ca_2 in Subcellular Domains and Intracellular Organelles." *Immunology and Cell Biology* 78 (4): 403–407.

Montgomery, Edmund. 1867. *On the Formation of So-Called Cells in Animal Bodies*. London: John Churchill and Sons.

Moore, John A. 1963. *Heredity and Development*. Oxford: Oxford University Press.

Moravec, Hans. 1988. *Mind Children: The Future of Robot and Human Intelligence*. Cambridge: Harvard University Press.

Morgan, Mary S., and Margaret C. Morrison, eds. 1999. *Models as Mediators: Perspectives on Natural and Social Science*. Cambridge: Cambridge University Press.

Morgan, T. H. 1924. "Mendelian Heredity in Relation to Cytology." In E. V. Cowdry, ed., *General Cytology*, pp. 691–734. Chicago: University of Chicago Press.

———. 1926a. *The Theory of the Gene*. New Haven: Yale University Press.

———. 1926b. "Genetics and the Physiology of Development." *American Naturalist* 60 (671): 489–515.

Morgan, T. H. 1934. *Embryology and Genetics.* New York: Columbia University Press.

Morrison, Margaret C. 1998. "Modelling Nature: Between Physics and the Physical World." *Philosophia Naturalis* 35: 65–85.

———. 1999. "Models as Autonomous Agents." In Morgan and Morrison 1999.

Muller, H. J. 1929. "The Gene as the Basis of Life." [Presented before the International Congress of Plant Sciences, Section of Genetics, Symposium on "The Gene," Ithaca, NY, August 19, 1926.] *Proceedings of the International Congress of Plant Sciences* 1: 897–921.

———. 1951. "The Development of the Gene Theory." In L. C. Dunn, ed., *Genetics in the Twentieth Century.* New York: Macmillan.

Mullins, M. C., M. Hammerschmidt, D. A. Kane, et al. 1996. "Genes Establishing Dorsoventral Pattern Formation in the Zebrafish Embryo: The Ventral Specifying Genes." *Development* 123: 81–93.

Murray, James D. 1988. "How the Leopard Gets Its Spots." *Scientific American* 258 (3): 80–87.

———. 1990. "Discussion: Turing's Theory of Morphogenesis—Its Influence on Modelling Biological Pattern and Form." *Bulletin of Mathematical Biology* 52 (1/2): 119–152.

Nanney, David L. 1957. "The Role of the Cytoplasm in Heredity." In William D. McElroy and Bentley Glass, eds., *The Chemical Basis of Heredity,* pp. 134–163. Baltimore: Johns Hopkins Press.

Needham, Joseph. "Biochemical Aspects of Form and Growth." In Whyte 1951, pp. 77–90.

Nijhout, H. F. 1985. "Independent Development of Homologous Pattern Elements in the Wing Patterns of Butterflies." *Developmental Biology* 108 (1): 146–151.

———. 1991. *The Development and Evolution of Butterfly Wing Patterns.* Washington, DC: Smithsonian Institution Press.

Niklas, Karl J. 1992. *Plant Biomechanics: An Engineering Approach to Plant Form and Function.* Chicago: University of Chicago Press.

Novick, Aaron, and Leo Szilard. 1954. "Experiments with the Chemostat on the Rates of Amino Acid Synthesis in Bacteria." In

Edgar J. Boell, ed., *Dynamics of Growth Processes*, pp. 21–32. Princeton: Princeton University Press.

Nüsslein-Volhard, C. 1991. "Determination of the Embryonic Axes of Drosophila." *Development* (supplement) 1: 1–10.

Odell, Garrett M., G. Oster, B. Burnside, and P. Alberch. 1981. "The Mechanical Basis of Morphogenesis. I: Epithelial Folding and Invagination." *Developmental Biology* 85: 446–462.

Odell, Garrett, George von Dassow, Eli Meir, and Ed Munro. 2001. "Robust Gene Networks Are the Only Networks Natural Selection Can Evolve." In preparation.

Olby, Robert. 1974. *The Path to the Double Helix*. Seattle: University of Washington Press.

———. 1986. "Structural and Dynamical Explanations in the World of Neglected Dimensions." In Horder et al. 1986, pp. 275–308.

Oparin, Aleksandr. 1953 [1938]. *Origin of Life*. New York: Dover.

Opitz, J. M. 1985. "The Developmental Field Concept." *American Journal of Medical Genetics* 21 (1): 1–11.

O'Rourke, N. A., and Scott E. Fraser. 1990. "Dynamic Changes in Optic Fiber Terminal Arbors Lead to Retinotopic Map Formation: An In Vivo Confocal Microscopic Study." *Neuron* 5: 159–171.

Orr-Weaver, T. L. 1995. "Meiosis in Drosophila: Seeing Is Believing." *Proceedings of the National Academy of Sciences USA* 92 (23): 10443–10449.

Oster, G., H. Wang, and M. Grabe. 2000. "How Fo-ATPase Generates Rotary Torque." *Philosophical Transactions of the Royal Society, London, B* 355: 523–528.

Oyama, Susan. 1985. *The Ontogeny of Information*. New York: Cambridge University Press.

Oyama, Susan, P. E. Griffiths, and R. D. Gray, eds. 2001. *Cycles of Contingency: Developmental Systems and Evolution*. Cambridge: MIT Press.

Palsson, Eirikur, and Hans G. Othmer. 2000. "A Model for Individual and Collective Cell Movement in *Dictyostelium discoideum*." *Proceedings of the National Academy of Sciences USA* 97: 10448–10453.

Parnes, Ohad. 2000. "The Envisioning of Cells." *Science in Context* 13 (1): 71–92.

Pattee, Howard H. 1989. "Simulations, Realizations, and Theories of Life." In Langton 1989, pp. 63–77.

Pawley, James B., ed. 1995. *Handbook of Biological Confocal Microscopy.* New York: Plenum Press.

Pearson, Karl. 1892. *Grammar of Science.* London: W. Scott.

Pinto-Correia, Clara. 1997. *The Ovary of Eve.* Chicago: University of Chicago Press.

Pirie, N. W. 1937. "The Meaninglessness of the Terms Life and Living." In Joseph Needham and David E. Green, eds., *Perspectives in Biochemistry,* pp. 11–22. Cambridge: Cambridge University Press.

Piston, David W. 1999. "Imaging Living Cells and Tissues by Two-Photon Excitation Microscopy." *Trends in Cell Biology* 9: 69–72.

Polanyi, Michael. 1958. *Personal Knowledge.* Chicago: University of Chicago Press.

Pratelle, Aristides. 1913. "Discussion by the French 'Plasmogenists' on the Origin of Life." *The Monist* 23: 617–623.

Prigogine, I., and R. Lefever. 1968. "Symmetry-Breaking Instabilities in Dissipative Systems, Part II." *Journal of Chemical Physics* 48: 1695.

Prigogine, I., R. Lefever, A. Goldbeter, and M. Herschkowitz-Kaufman. 1969. "Symmetry Breaking Instabilities in Biological Systems." *Nature* 223 (209): 913–916.

Prigogine, I., and G. Nicolis. 1967. "On Symmetry-Breaking Instabilities in Dissipative Systems." *Journal of Chemical Physics* 46: 3542.

Prochiantz, Alain. 2001. *Machine-esprit.* Paris: Editions Odile Jacob.

Provine, W. B. 1971. *The Origins of Theoretical Population Genetics.* Chicago: University of Chicago Press.

Rader, K. A. 1998. "'The Mouse People': Murine Genetics at the Bussey Institution, 1909–1936." *Journal of the History of Biology* 31: 327–354.

Ramón y Cajal, Santiago. 1996 [1937]. *Recollections of My Life.* E. Horne Craigie with Juan Cano, trans. Cambridge: MIT Press.

Rashevsky, Nicolas. 1933. "The Theoretical Physics of the Cell as a Basis for a General Physico-Chemical Theory of Organic Form." *Protoplasma* 20: 180–188.

———. 1934. "Physico-Mathematical Aspects of Cellular Multiplica-

tion and Development." *Cold Spring Harbor Symposia for Quantitative Biology* 2: 188–198.
———. 1940. "An Approach to the Mathematical Biophysics of Biological Self-Organization and of Cell Polarity." *Bulletin of Mathematical Biophysics* 2: 15–25.
Rasmussen, Nicholas. 1997. *Picture Control: The Electron Microscope and the Transformation of Biology in America, 1940–1960.* Stanford: Stanford University Press.
Ravin, Arnold. 1977. "The Gene as Catalyst; The Gene as Organism." *Studies in the History of Biology* 1: 1–45.
Ray, Thomas S. 1995. "An Evolutionary Approach to Synthetic Biology." In C. G. Langton, ed., *Artificial Life: An Overview,* pp. 179–207. Cambridge: MIT Press.
———. 1998. "Selecting Naturally for Differentiation: Preliminary Evolutionary Results." *Complexity* 3 (5): 25–33.
Rayleigh, John William Strutt. 1916. "On Convection Currents in a Horizontal Layer of Fluid, When the Higher Temperature Is on the Under Side." *Philosphical Magazine* 32: 529–546.
Resnick, Mitchel. 1994. *Turtles, Termites, and Traffic Jams: Explorations in Massively Parallel Microworlds.* Cambridge: MIT Press.
Reznikoff, William S. 1972. "The Operon Revisited." *Annual Review of Genetics* 6: 133–156.
Ritterbush, P. C. 1968. *The Art of Organic Form.* Washington, DC: Smithsonian Institution Press.
Roe, Shirley A. 1981. *Matter, Life, and Generation.* Cambridge: Cambridge University Press.
Roger, Jacques. 1997 [1963]. *The Life Sciences in Eighteenth Century France,* ed. Keith R. Benson, trans. Robert Ellrich. Stanford: Stanford University Press.
Rohrlich, Fritz. 1991. "Computer Simulation in the Physical Sciences." *Philosophy of Science Association 1990* 2: 507–518.
Rosen, Robert. 1970. *Dynamical Theory in Biology.* New York: Wiley-Interscience.
———. 1972. "Nicolas Rashevsky 1899–1972." In *Progress in Theoretical Biology* 2: xi–xiv.
Rosenberg, Alex. 1997. "Reductionism Redux: Computing the Embryo." *Biology and Philosophy* 12: 445–470.

Rosenbleuth, A., N. Wiener, and J. Bigelow. 1943. "Behavior, Purpose, and Teleology." *Philosophy of Science* 10: 18–24.

Roth, S., D. Stein, and C. Nüsslein-Volhard. 1989. "A Gradient of Nuclear Localization of the Dorsal Protein Determines Dorsoventral Pattern in the Drosophila Embryo." *Cell* 59 (6): 1189–1202.

Rothman, Daniel H., and S. Zaleski. 1997. *Lattice-Gas Cellular Automata: Simple Models of Complex Hydrodynamics.* Cambridge: Cambridge University Press.

Salazar-Ciudad, I., S. A. Newman, and R. V. Solé. 2001a. "Phenotypic and Dynamical Transitions in Model Genetic Networks. I. Emergence of Patterns and Genotype-Phenotype Relationships." *Evolution and Development* 3 (2): 84–94.

Salazar-Ciudad, I., R. V. Solé, and S. A. Newman. 2001b. "Phenotypic and Dynamical Transitions in Model Genetic Networks. II. Application to the Evolution of Segmentation Mechanisms." *Evolution and Development* 3 (2): 95–103.

Sapp, Jan. 1987. *Beyond the Gene: Cytoplasmic Inheritance and the Struggle for Authority in Genetics.* New York: Oxford University Press.

Saunders, Peter. 1993. "Alan Turing and Biology." *IEEE Annals of the History of Computing* 15 (3): 33–36.

Saunders, Peter, ed. 1992. *Morphogenesis: Collected Works of A. M. Turing,* vol. 3. Amsterdam: North-Holland.

Schäfer, Sir Edward Albert. 1912. "The Nature, Origin, and Maintenance of Life." *Nature* 90: 7–19.

Schaffer, Simon. 1997. "Forgers and Authors in the Baroque Economy." Presented at the conference on "What Is a Scientific Author?" Harvard University, March 7–9.

Schiller, Joseph. 1978. *La notion d'organisation dans l'histoire de la biologie.* Paris: Maloine.

Service, Robert F. 1999. "Watching DNA at Work." *Science* 283 (5408): 1668–1669.

Shapin, Steven, and Simon Schaffer. 1985. *Leviathan and the Air-Pump: Hobbes, Boyle, and the Experimental Life.* Princeton: Princeton University Press.

Sibum, H. Otto. 1998. "Les gestes de la mesure: Joule, les pratiques de la brasserie et la science." *Annales: économies, sociétés, civilisations* 53: 745–774.

Siegel, Daniel. 1992. *Innovation in Maxwell's Electromagnetic Theory: Molecular Vortices, Displacement Current, and Light.* Cambridge: Cambridge University Press.

Sims, Karl. 1991. "Interactive Evolution of Dynamical Systems." In Francisco J. Varela and Paul Bourgine, eds., *Towards a Practice of Autonomous Systems: Proceedings of the First European Conference on Artificial Life,* pp. 171–178. Cambridge: MIT Press.

Smith, James C. 2000. "From Engineering to Positional Information to Public Understanding: An Interview with Lewis Wolpert." *International Journal of Developmental Biology* 44: 85–91.

Spengler, Sylvia J. 2000. "Techview: Computers and Biology: Bioinformatics in the Information Age." *Science* 287 (5456): 1221–1223.

Stent, Gunther. 1968. "That Was the Molecular Biology That Was." *Science* 160: 390–395.

Stern, Curt. 1955. "Gene Action." In Benjamin Willier, Paul Weiss, and Viktor Hamburger, eds., *Analysis of Development,* pp. 151–169. Philadelphia: W. B. Saunders.

Stewart, Ian. 1998. *Life's Other Secret: The New Mathematics of the Living World.* London: Allen Lane.

Strick, James. 1999. "Darwinism and the Origin of Life: The Role of H. C. Bastian in the British Spontaneous Generation Debates, 1868–1873." *Journal of the History of Biology* 32 (1): 51–92.

Stumpf, Hildegard. 1966. "Mechanism by Which Cells Estimate Their Location within the Body." *Nature* 212: 430–431.

Sturtevant, Alfred H. 1932. "The Use of Mosaics in the Study of the Developmental Effects of Genes." In Donald F. Jones, ed., *Proceedings of the Sixth International Congress of Genetics,* p. 304. Menasha, WI: Brooklyn Botanic Garden.

Sugita, Motoyosi. 1961a. "The Switching Circuit Logically or Functionally Equivalent to a System of Biochemical Reactions." *Journal of the Physical Society of Japan* 16 (4): 737–740.

———. 1961b. "Functional Analysis of Chemical Systems *In Vivo* Using a Logical Circuit Equivalent." *Journal of Theoretical Biology* 1: 415–430.

———. 1975. "Functional Analysis of Chemical Systems In Vivo Using a Logical Circuit Equivalent. V. Molecular Biological Interpretation of the Self-Reproducing Automata Theory and

Chemico-Physical Interpretation of Information in Biological Systems." *Journal of Theoretical Biology* 53 (1): 223–237.

Szathmáry, Eörs. 2001. "Evolution: Developmental Circuits Rewired." *Nature* 411 (6834): 143–144.

Szent-Gyorgyi, Albert. 1972. *What Is Life?* Del Mar, CA: CRM Books.

Szilard, Leo. 1960. "The Control of the Formation of Specific Proteins in Bacteria and in Animal Cells." *Proceedings of the National Academy of Sciences USA* 46 (3): 277–292.

Thieffry, D. 1999. "From Global Expression Data to Gene Networks." *Bioessays* 21 (11): 895–899.

Thomas, René. 1973. "Boolean Formalization of Genetic Control Circuits." *Journal of Theoretical Biology* 42 (3): 563–585.

Thomas, René, and Richard D'Ari. 1990. *Biological Feedback.* Boca Raton: CRC Press.

Thompson, D'Arcy Wentworth. 1915. "Morphology and Mathematics." *Transactions of the Royal Soc. Edinburgh* 1, pt. 4 (27): 857–895.

———. 1917. *On Growth and Form.* Cambridge: Cambridge University Press.

———. 1942. *On Growth and Form.* 2nd ed. Cambridge: Cambridge University Press.

Thompson, Ruth D'Arcy. 1958. *D'Arcy Wentworth Thompson.* London: Oxford University Press.

Thomson, J. Arthur. 1917. "Foundations of Bio-Physics." *Nature* 100 (2498): 21–22.

Tiles, Mary, and Jim Tiles. 1993. *An Introduction to Historical Epistemology: The Authority of Knowledge.* Oxford: Blackwell.

Toffoli, T. 1984. "Cellular Automata as an Alternative to (Rather than an Approximation of) Differential Equations in Modeling Physics." *Physica* 10D: 117–127.

Toffoli, T., and N. Margolus. 1987. *Cellular Automata Machines: A New Environment for Modeling.* Cambridge: MIT Press.

Trevinarus, G. R. 1802. *Biologie oder Philosophie der lebenden Natur,* Bd. 1. Göttingen.

Troland, Leonard Thompson. 1914. "The Chemical Origin and Regulation of Life." *The Monist* 24: 92–133.

Tufte, Edward R. 1997. *Visual Explanations: Images and Quantities, Evidence and Narrative.* Cheshire, CT: Graphics Press.

Turing, A. M. 1952. "The Chemical Basis of Morphogenesis." *Philosophical Transactions of the Royal Society, London, B* 237: 37–72.

Turney, Jon. 1998. *Frankenstein's Footsteps: Science, Genetics and Popular Culture.* New Haven: Yale University Press.

Tyndall, J. 1870. *Essays on the Use and Limits of the Imagination in Science.* London: Longman and Green.

Umbarger, H. E. 1956. "Evidence for a Negative Feedback Mechanism in the Biosynthesis of Isoleucine." *Science* 123: 848.

Vaihinger, Hans. 1924. *The Philosophy of "As If."* C. K. Ogden, trans. New York, London: Harcourt, Brace.

Vance, Stanley. 1960. *Management Decision Simulation.* New York: McGraw-Hill.

Verlet, Loup. 1993. *La Malle de Newton.* Paris: Éditions Gallimard.

———. 1996. "'F = MA' and the Newtonian Revolution: An Exit from Religion through Religion." *History of Science* 34: 303–346.

Vichniac, G. Y. 1984. "Simulating Physics with Cellular Automata." *Physica* 10D: 96–116.

Von Dassow, George, Eli Meir, Edwin H. Munro, and Garrett M. Odell. 2000. "The Segment Polarity Network Is a Robust Developmental Module." *Nature* 406: 188–192.

Von Neumann, John. 1966. *Theory of Self-Reproducing Automata.* Arthur Burks, ed. Urbana: University of Illinois Press.

Vries, Hugo de. 1889. *Intracellular Pangenesis.* Trans. 1910. Chicago: Open Court.

Waddington, C. H. 1940. *Organisers and Genes.* Cambridge: Cambridge University Press.

———. 1941. "Canalization of Development and the Inheritance of Acquired Characteristics." *Nature* 150: 563–565.

———. 1953. "The Origin of Biological Pattern." *New Biology* 15: 117–125.

———. 1954. "The Cell Physiology of Early Development." In Kitching 1954, pp. 105–119.

———. 1957. *The Strategy of the Genes.* London: Allen and Unwin.

———. 1961. *The Nature of Life.* New York: Harper Torchbooks.

———. 1962. *New Patterns in Genetics and Development.* New York: Columbia University Press.

Wagner, Andreas. 2000. "Robustness against Mutations in Genetic Networks of Yeast." *Nature Genetics* 24 (4): 355–361.

Wainwright, T., and B. J. Alder. 1958. "Molecular Dynamics Computations for the Hard Sphere System." *Nuovo Cimento* (supplement) 9, series 10: 116–143.

Waldrop, J. Mitchell. 1992. *Complexity.* New York: Simon and Schuster.

Warburton, Frederick E. 1955. "Feedback in Development and Its Evolutionary Significance." *American Naturalist* 89: 129–140.

Weinberg, Robert A. 1985. "The Molecules of Life." *Scientific American* 253 (4): 48–57.

Weinberg, Steven. 2001. "Can Science Explain Everything? Anything?" *New York Review of Books,* May 31.

Weininger, Stephen J. 1998. "Contemplating the Finger: Visuality and the Semiotics of Chemistry." *HYLE—An International Journal for the Philosophy of Chemistry* 4 (1): 3–27.

Weismann, August. 1889. *Continuity of the Germ Plasm.* In E. Poulton et al., eds., *Essays upon Heredity and Kindred Biological Problems.* Oxford: Clarendon Press; originally published in German in 1885.

Weiss, Ron, George Homsy, and Thomas F. Knight. 1999. "Toward *In Vivo* Digital Circuits." Presented at Dimacs Workshop on Evolution as Computation, Princeton, NJ. Available at: http://www.swiss.ai.mit.edu/projects/amorphous/paperlisting.html#invivo-circuits.

Wheeler, William Morton. 1929. "Present Tendencies in Biological Theory." *Scientific Monthly* 28 (2): 97–109.

Whitaker, Michael. 2000. "Fluorescent Tags of Protein Function in Living Cells." *BioEssays* 22 (2): 180–187.

Whyte, Lancelot L., ed. 1951. *Aspects of Form: A Symposium on Form in Nature and Art.* London: Percy Lund Humphries.

Wiener, Norbert. 1948. *Cybernetics: or, Control and Communication in the Animal and Machine.* Cambridge: MIT Press.

———. 1950. *Human Use of Human Beings: Cybernetics and Society.* Boston: Houghton Mifflin.

Wilcox, M., G. J. Mitchison, and R. J. Smith. 1973. "Pattern Formation in the Blue-Green Alga, Anabaena. I. Basic Mechanisms." *Journal of Cell Science* 12 (3): 707–723.

Wilson, Catherine. 1995. *The Invisible World: Early Modern Philosophy and the Invention of the Microscope.* Princeton: Princeton University Press.

Wilson, E. B. 1925. *The Cell in Development and Heredity.* 3rd ed. New York: Macmillan.

———. 1934. "Mathematics of Growth." *Cold Spring Harbor Symposia for Quantitative Biology* 2: 199–202.

Winsberg, Eric. 1999. "Sanctioning Models: The Epistemology of Simulation." *Science in Context* 12 (2): 275–292.

Wise, Norton. 1977. "The Mutual Embrace of Electricity and Magnetism." *Science* 203: 1310–1318.

Wolfram, Stephen, ed. 1986. *Theory and Applications of Cellular Automata.* Singapore: World Scientific Publishing Co.

Wolpert, Lewis. 1968. "The French Flag Problem: A Contribution to the Discussion on Pattern Development and Regulation." *Towards a Theoretical Biology* 1: 125–133.

———. 1969. "Positional Information and the Spatial Pattern of Cellular Differentiation." *Journal of Theoretical Biology* 25: 1–47.

———. 1970. "Positional Information and Pattern Formation." *Towards a Theoretical Biology* 3: 198–230.

———. 1971. "Positional Information and Pattern Formation." *Current Topics in Developmental Biology* 6: 183–224.

———. 1972. "The Concept of Positional Information and Pattern Formation." *Towards a Theoretical Biology* 4: 83–94.

———. 1975. "Positional Information and Pattern Formation in Hydra." *Journal of Embryology and Experimental Morphology* 33 (2): 511–521.

———. 1978. "Pattern Formation in Biological Development." *Scientific American* 239: 154–164.

———. 1981a. "Pattern Formation in Limb Morphogenesis." In J. W. Sauer, ed., *Progress in Developmental Biology,* pp. 141–152. New York: G. Fischer.

———. 1981b. "Positional Information and Pattern Formation." *Philosophical Transactions of the Royal Society, London, B* 295: 441–450.

———. 1985. "Positional Information and Pattern Formation." In G. Edelman, ed., *Molecular Determinants of Animal Form,* pp. 423–433. New York: A. R. Liss.

———. 1986. "Gradients, Position and Pattern: A History." In T. J.

Horder, J. A. Witkowski, and C. C. Wylie, eds., *History of Embryology,* pp. 347–362. Cambridge: Cambridge University Press.
———. 1989. "Positional Information Revisited." *Development* (supplement), pp. 3–12.
———. 1994a. "Positional Information and Pattern Formation in Development." *Development Genetics* 15: 485–490.
———. 1994b. "Do We Understand Development?" *Science* 266: 571–572.
———. 1995. "Bet on Positional Information." *Nature* 373: 112.
———. 1996. "One Hundred Years of Positional Information." *Trends in Genetics* 12 (9): 359–364.
Wolpert, Lewis (with Rosa Beddington, Jeremy Brockes, Thomas Jessell, Peter Lawrence, and Elliot Meyerowitz). 1996. *Principles of Development.* Oxford: Oxford University Press.
Wolpert, Lewis, and T. Gustafson. 1961. "Studies on the Cellular Basis of Morphogenesis of the Sea Urchin Embryo: Development of the Skeletal Pattern." *Experimental Cell Research* 25: 311–325.
Wolpert, Lewis, and J. H. Lewis. 1975. "Towards a Theory of Development." *Federation Proceedings* 34 (1): 14–20.
Wolpert, Lewis, and Wilfred D. Stein. 1984. "Positional Information and Pattern Formation." In G. Malacinski and S. Bryant, eds., *Pattern Formation: A Primer in Developmental Biology,* pp. 3–21. New York: Macmillan.
Woodward, James. 1997. "Explanation, Invariance, and Intervention." *Philosophy of Science* 64 (Proceedings): S26–S41.
———. 2001. *A Theory of Explanation.* In preparation.
Wray, Gregory A. 1998. "Promoter Logic." *Science* 279: 1871–1872.
Wright, Sewall. 1945. "Genes as Physiological Agents." *American Naturalist* 79 (783): 289–303.
Yates, Richard A., and Arthur Pardee. 1956. "Control of Pyrimidine Biosynthesis in *Escherichia coli* by a Feed-Back Mechanism." *Journal of Biological Chemistry* 221: 757–769.
You, H. X., and L. Yu. 1999. "Atomic Force Microscopy Imaging of Living Cells: Progress, Problems and Prospects." *Methods Cell Science* 21 (1): 1–17.
Yuh, Chiou-Hwa, H. Bolouri, and E. H. Davidson. 1998. "Genomic cis-Regulatory Logic: Experimental and Computational Analysis of a Sea Urchin Gene." *Science* 279: 1896–1902.

———. 2001. "Cis-Regulatory Logic in the *Endo16* Gene: Switching from a Specification to a Differentiation Mode of Control." *Development* 128: 617–629.

Yuh, Chiou-Hwa, and E. H. Davidson. 1996. "Modular cis-Regulatory Organization of *Endo16*." *Development* 122: 1069–1082.

Yuh, Chiou-Hwa, J. G. Moore, and E. H. Davidson. 1996. "Quantitative Functional Interrelations within the cis-Regulatory System of the S. purpuratus *Endo16* Gene." *Development* 122: 4045–4056.

Yuh, Chiou-Hwa, A. Ransick, P. Martinez, R. J. Britten, and E. H. Davidson. 1994. "Complexity and Organization of DNA-Protein Interactions in the 5′ Regulatory Region of an Endoderm-Specific Marker Gene in the Sea Urchin Embryo." *Mechanisms of Development* 47: 165–186.

Zwanziger, Lee L. 1989. "Structure and Modeling in Biological Theories of Embryological Pattern Formation, 1968–1986." Ph.D. diss., University of Pittsburgh.

Index